Basic Concepts of Linear Order

Combinatorics for Computer Science (Unit 1)

S. Gill Williamson

Preface

From 1972 to 1990 I ran a graduate seminar on algorithmic combinatorics in the Department of Mathematics, UCSD. Over time, I developed a series of notes or "units of study" to prepare beginning graduate students for this seminar. In 1985, these units of study were combined and published as a book, *Combinatorics for Computer Science (CCS)*, published by Computer Science Press. Each of the units of study became a chapter in this book.

My general goal is to re-create the original presentation of these (largely independent) units in a form that is convenient for individual selection and study. Here, we isolate Unit 1, corresponding to Chapter 1 of *CCS*, and reconstruct the original very helpful unit specific index associated with this unit.

Theorems, figures, examples, etc., are numbered sequentially: EXERCISE 1.38 and FIGURE 1.62 refer to numbered items 38 and 62 of Unit 1 (or Chapter 1 in *CCS*), etc. *CCS* contains an extensive bibliography for work prior to 1985. For further references and ongoing research, search the Web, particularly Wikipedia and the mathematics arXiv (arXiv.org).

These notes focus on the visualization of algorithms through the use of graphical and pictorial methods. This approach is both fun and powerful, preparing you to invent your own algorithms for a wide range of problems.

S. Gill Williamson, 2012
http : \www.cse.ucsd.edu\ ∼ gill

Table of Contents for Unit 1

UNIT 1

Basic Concepts of Linear Order

We shall be concerned with studying basic finite sets from a point of view that facilitates computations with these sets. Permutations, subsets, graphs, tree structures, partially and totally ordered sets, etc., are interesting in their own right. They are also the fundamental building blocks for describing and constructing a variety of algorithms. Our point of view will be motivated largely by these algorithmic considerations. The construction and manipulation of linear lists is one of the most fundamental techniques in the design and analysis of algorithms. We begin by looking at the idea of a *linear order* from a somewhat "formal" point of view. In the process we develop some basic ideas to be used later on.

1.1 DEFINITION.

Let S be a set. A relation on S is a function from the Cartesian product, $S \times S$ to any set T with two elements.

For example, $T = \{0,1\}$, $T = \{-1,+1\}$, and $T = \{\text{false, true}\}$ could all be used as the range T of a relation ρ on S. If the set T is fixed and S is finite, then it is easily seen that there are 2^p, $p = |S|^2$, functions from $S \times S$ to T (if S is a finite set, $|S|$ denotes the *cardinality* or *number of elements* of S). The notion of a relation is, of course, a triviality in full generality. Two special classes of relations, however, play an important descriptive role in the study of algorithms. For the next definition, let $\rho : S \times S \to T$ be a relation. Fix T = {false, true}. We shall use the equivalent but more suggestive notation "x ρ y" for "$\rho(x,y) = $ true" and "x $\rho\!\!\!/$ y" for "$\rho(x,y) = $ false."

1.2 DEFINITION.

A relation ρ on S is

(1) *Reflexive* if for all $x \in S$, $x \rho x$.
(2) *Symmetric* if for all $x, y \in S$, $x \rho y$ implies $y \rho x$.
(2') *Antisymmetric* if for all $x, y \in S$, $x \rho y$, and $y \rho x$ implies $x = y$.
(3) *Transitive* if for all $x, y, z \in S$, $x \rho y$ and $y \rho z$ implies $x \rho z$.

A relation ρ that satisfies 1, 2, and 3 is called an *equivalence relation*. A relation ρ that satisfies 1, 2', and 3 is called an *order relation*.

3

The general structure of equivalence relations is easy to understand because of the well-known correspondence between equivalence relations and partitions of a set.

1.3 DEFINITION.

Let S be a set. A *partition* of S is a collection $\mathscr{C}$ of subsets of S such that $\bigcup_{A \in \mathscr{C}} A = S$, and if A and B are elements of $\mathscr{C}$, then either A = B or A and B are disjoint. The elements of $\mathscr{C}$ (which are subsets of S by definition) are called the *blocks* of $\mathscr{C}$. $\mathscr{C}$ is *discrete* if each block has one element. The empty set $\phi \notin \mathscr{C}$.

Thus, if N^+ is the set of positive integers, then $\mathscr{C} = \{E, O\}$, where E is the set of even numbers and O is the set of odd numbers in N^+, is a partition of N^+. The collection $\{\{1,3,7\}, \{2,4,5,6\}, \{8\}\}$ is a partition of $S = \{1, \ldots, 8\}$.

1.4 DEFINITION.

Let ρ be an equivalence relation on S. For each $s \in S$, let E_s be the set $\{x: x \in S, x \rho s\}$. The set E_s is called the *equivalence class of s with respect to ρ or the equivalence class of s*.

1.5 THEOREM.

If ρ is an equivalence relation on a set S, then the collection $\mathscr{C} = \{E_s: s \in S\}$ of all ρ equivalence classes is a partition of S. Conversely, if $\mathscr{C}$ is any partition of S, and x and y are elements of S, then define $x \rho y$ if x and y belong to the same block of $\mathscr{C}$. Then ρ is an equivalence relation on S, and $\mathscr{C}$ is the collection of equivalence classes.

THEOREM 1.5 finds its way into many undergraduate courses in mathematics (discrete math, real analysis, logic, algebra, group theory, linear algebra). The reader should attempt to reconstruct the proof and consult a reference if necessary. Although the general structure of equivalence relations is quite clear from THEOREM 1.5, particular relations might not obviously be equivalence relations at first glance. Transitivity, in particular, is sometimes a bit tricky to verify. Once the axioms are verified for ρ, then the partition into equivalence classes follows from THEOREM 1.5.

1.6 NOTATION.

The set of integers $1, \ldots, n$ will be denoted by $\underline{n}$. If A and B are sets, we write f: $A \rightarrow B$ for a function f with domain A and range B. The set of all such functions will be denoted by B^A (note that if A and B are finite with cardinality $|A| = a$ and $|B| = b$, then $|B^A| = b^a$, hence the notation). The *Image*(f) is the set $\{f(a): a \in A\}$. For each $b \in B$, $f^{-1}(b)$ is called the "inverse image of b" and is the set $\{a: a \in A, f(a) = b\}$. The collection $\mathscr{C} = \{f^{-1}(b): b \in Image(f)\}$

is a partition of A and is called the Coimage(f). If Image(f) = B, then f is a *surjection*. If Coimage(f) is discrete, then f is an *injection*. If f is both an injection and a surjection, it is a *bijection*. It is easily seen that if |A| = |B| (both finite), then f is an injection if and only if it is a surjection. The injections of A^A are called the *permutations* of A, A finite.

As an example of the above ideas, consider $f \in \underline{6}^{\underline{6}}$ where $f = \begin{pmatrix} 1\ 2\ 3\ 4\ 5\ 6 \\ 1\ 3\ 1\ 3\ 2\ 2 \end{pmatrix}$. The Image(f) = {1,2,3}. Coimage(f) = {{1,3}, {5,6}, {2,4}}.

This function can be written in *one-line notation* as (1 3 1 3 2 2). This notation specifies the function if the domain is known and specified in some order. Some permutations of $\underline{6}$ in one-line notation are (5 6 3 2 1 4) or (4 2 3 5 6 1).

We shall see many examples in the text of equivalence relations. We mention a few examples here (it is conventional to use a symbol such as ~ rather than ρ when working with equivalence relations).

1.7 EXAMPLES OF EQUIVALENCE RELATIONS.

(1) S = $\underline{2}^{\underline{3}}$ with f ~ g if Image(f) = Image(g). The equivalence classes are {(1 1 1)}, {(1 1 2), (1 2 1), (2 1 1), (1 2 2), (2 1 2), (2 2 1)}, and {(2 2 2)}. We use one line notation for all functions.

(2) S = $\underline{2}^{\underline{3}}$ with f ~ g if Coimage(f) = Coimage(g). The equivalence classes are {(1 1 1), (2 2 2)}, {(1 1 2), (2 2 1)}, {(1 2 1), (2 1 2)}, and {(1 2 2), (2 1 1)}.

(3) S = $\underline{2}^{\underline{3}}$ with f ~ g if Max(f) = Max(g). There are two equivalence classes.

(4) S = $\underline{2}^{\underline{3}}$ with f ~ g if f is a cyclic shift of g: (1 1 2),(2 1 1), and (1 2 1) are cyclic shifts of each other.

(5) S = $\underline{2}^{\underline{3}}$ with f ~ g if f(1) + f(2) + f(3) = g(1) + g(2) + g(3).

For the reader who knows a little graph theory:

(6) Let G = (V,E) be a graph. Define an equivalence relation on V by x ~ y if there is a path in G from x to y. The equivalence classes are used to define the connected components of G.

(7) Let G = (V,E) be a graph. Define an equivalence relation on E by e ~ f if e and f lie on the same simple (not self-intersecting) cycle of G. The equivalence classes are used to define the biconnected components of G. Check transitivity here. Assume e ~ e.

(8) Let S = {(a,b): a and b integers, b ≠ 0}. Define (a,b) ~ (a′,b′) if ab′ = ba′. This is the equivalence relation used in the formal definition of the rational numbers.

(9) Let {a_n} and {b_n} be two infinite sequences of rational numbers. Define {a_n} ~ {b_n} if $\lim_{n \to \infty} (a_n - b_n) = 0$. Such an equivalence relation is used in the formal development of the real number system.

The general structure of order relations is more complex than that of equivalence relations. The notation $x \leq y$ or $x \mathrel{\underline{\propto}} y$ is often used for order relations rather than $x \rho y$. A set S together with an order relation $\mathrel{\underline{\propto}}$ is often called a "partially ordered set" or "poset." We write $(S, \mathrel{\underline{\propto}})$ to designate such a poset. We use $x \propto y$ to mean $x \mathrel{\underline{\propto}} y$, but $x \neq y$.

1.8 DEFINITION.

Let $(S, \mathrel{\underline{\propto}})$ be an ordered set. We say y *covers* x if $x \propto y$ and if, for all $z \in S$, $x \mathrel{\underline{\propto}} z \mathrel{\underline{\propto}} y$ implies $x = z$ or $y = z$. If y covers x, we write $x \overset{x}{c} y$ and we say that y is a successor of x and x a predecessor of y.

1.9 EXAMPLES OF ORDERED SETS.

(1) $(R, \leq)$ Real numbers with usual ordering.
(2) $(N, \leq)$ Positive integers with usual ordering.
(3) Given a set A, let $S = \mathscr{P}(A)$, the set of all subsets of A. Then $(S, \subseteq)$ is a poset where "$\subseteq$" denotes the usual set inclusion.
(4) $(\pi(A), \mathrel{\underline{\propto}})$, where $\pi(A) = $ partitions of a set A, and $\pi_2 \mathrel{\underline{\propto}} \pi_1$ if π_2 is a refinement of π_1, e.g., if $A = \underline{8}$, $\pi_1 = \{\{1,3,5\}, \{2,4,6\}, \{7,8\}\}$ and $\pi_2 = \{\{1,3\}, \{5\}, \{2,6\}, \{4\}, \{7,8\}\}$, then $\pi_2 \mathrel{\underline{\propto}} \pi_1$. (Blocks of π_1 are split further to get blocks of π_2).
(5) $(N, \mathrel{\underline{\propto}})$, where $x \mathrel{\underline{\propto}} y$ if and only if $x|y$ ("x divides y").
(6) $\underline{r}^{\underline{d}}$ (set of functions from $\underline{d}$ to $\underline{r}$), with $f \mathrel{\underline{\propto}} g$ if and only if $f(i) \leq g(i)$ for all i.

1.10 DEFINITION.

Given a poset $P = (S, \leq)$, let $P_{cov} = (S, \overset{x}{c})$. We create a diagram of P_{cov} by connecting a to b with a line *if and only if* a covers b. This diagram of P_{cov} is called the *Hasse diagram* of P.

1.11 HASSE DIAGRAM FOR S = $\underline{12}$, WITH x|y AS THE ORDER RELATION.

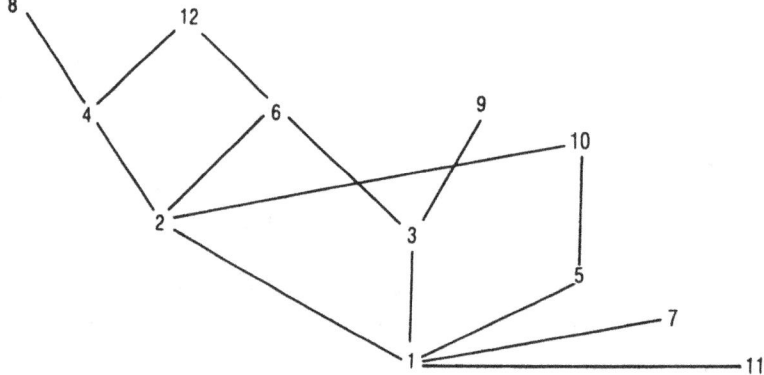

Figure 1.11

6

1.12 HASSE DIAGRAMS OF $(\mathscr{P}(\underline{2}), \subseteq)$, ALL SUBSETS OF $\underline{2}$ WITH INCLUSION AND $(\underline{5}, \leq)$.

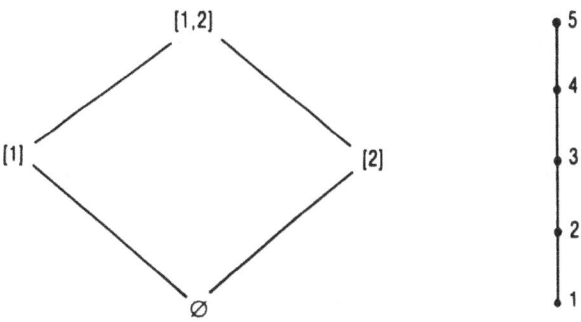

Figure 1.12

1.13 EXERCISE.

(1) Let $\mathscr{M}_n$ be the set of all $n \times n$ matrices with real entries. Define a relation $\sim$ on $\mathscr{M}_n \times \mathscr{M}_n$ by $(A,B) \sim (C,D)$ if $A + D = B + C$. Show that $\sim$ is an equivalence relation on $\mathscr{M}_n \times \mathscr{M}_n$.

(2) Let $\mathscr{M}_n$ be as in (1). Define $A \sim B$ if there is a nonsingular matrix P such that $P A P^{-1} = B$. Show that $\sim$ is an equivalence relation on $\mathscr{M}_n$. A student who has studied *linear algebra* should be able to give a good description of the equivalence classes for this equivalence relation.

(3) Let ρ_1 be a relation on S and ρ_2 a relation on T. Define a relation ρ_3 on $S \times T$ by $(s,t)\rho_3(s',t')$ if $s\rho_1s'$ *and* $t\rho_2t'$. Show that ρ_3 is an equivalence relation if both ρ_1 and ρ_2 are equivalence relations. Show ρ_3 is an order relation if both ρ_1 and ρ_2 are order relations.

(4) Construct the Hasse diagram of $(\mathscr{P}(\underline{4}), \subseteq)$ and $(\underline{18},|)$. See EXAMPLES 1.9(3) and 1.9(5). See Figure 1.11 for $(\underline{12},|)$.

(5) Let ρ be a reflexive, symmetric relation on S. Define a relation ρ' on S by $s\rho't$ if there exists some sequence $u_1,\ldots,u_p$ in S such that $s\rho u_1, u_1\rho u_2,\ldots,u_p\rho t$. Show that ρ' is an equivalence relation on S.

(6) Referring to EXAMPLES 1.9(3) and 1.9(6), consider $(\mathscr{P}(\underline{d}), \subseteq)$ and $(\{0,1\}^{\underline{d}}, \leq)$. For $A \in \mathscr{P}(\underline{d})$, let $f_A \in \{0,1\}^{\underline{d}}$ be defined by $f_A(x) = 0$ if $x \notin A$, 1 if $x \in A$. Show that the map $\varphi(A) = f_A$ is a bijection from $\mathscr{P}(\underline{d})$ to $\{0,1\}^{\underline{d}}$. Show that $A \subseteq B$ if and only if $f_A \leq f_B$. Such a map φ is called an *order preserving bijection* between the two posets. The function f_A is called the *characteristic function* of A.

There is an extensive and quite fascinating mathematical theory of partially ordered sets. From an algorithmic point of view, however, the most basic techniques involve working with various types of *linear orders*. A poset $(S, \leq)$ is *linearly ordered* if for every x and y in S either $x \le y$ or $y \le x$. The Hasse diagram of a linearly ordered set is a chain (as, for example, $(\underline{5}, \le)$ of Figure 1.12). The purely mathematical idea of a linear order still permits many other digressions. Fermat's conjecture states that for positive integers a,b,c and $n > 2$, $a^n + b^n \ne c^n$. We could let $S = \{x,y\}$ and say that $x < y$ if Fermat's conjecture is true and $y < x$ if Fermat's conjecture is false. If Fermat's conjecture is either true or false, then this linear order is well defined. If we could resolve Fermat's conjecture, the linear order would be well defined but totally uninteresting *as a linear order*. In the rest of this chapter, we take a naive point of view towards linear orders and simply explore those aspects that have been shown by experience to relate to the study of algorithms.

It is important to consider how in practice a linear order on a set S might be specified. One way is to list the objects of the set in some sort of "linear array." Consider ¢, %, &, \$, *, #, (,@,), + which specifies a linear order on a standard set of text symbols. Is $\$ < \#$? One way to decide this is to read the array from first to last (left to right here) and check which of the two symbols being compared comes first. In some cases, this might be the only way to decide which of two elements in a linearly ordered set is the smaller. Consider the linearly ordered set @, ¢, x, z, #. The number of straight line segments in these symbols is respectively 0, 1, 2, 3, and 4. Thus, associated with each symbol in the list is an intrinsic method for computing its position in the list. This idea can be important when deciding which of two elements should appear first in long lists. Even if the correspondence is not bijective (unique for each symbol), such an intrinsic association might be helpful. Consider the list @, &, +, x, v, y, z, N, #. Here, the straight line segment computation yields 0, 0, 2, 2, 2, 2, 3, 3, 4. Consider Table 1.14, as follows.

1.14 HASHING TABLE.

0:	@ &
1:	empty
2:	+ x v y
3:	z N
4:	#

In order to compare two elements in the list one might first compute the number of straight line segments in the two symbols. For + one would compute 2, and for N one would compute 3. If one can then recognize easily that 2 is less than 3, then one can conclude that $+ < N$. If one is required to compare + and v, then one computes 2 for both symbols. In this case it is necessary to go to the line labeled 2 in TABLE 1.14. One then can look at the sublist of symbols in line 2 to decide if $+ < v$. One can imagine large-scale applications of this sort of idea (called "hashing" by computer scientists). Many interesting questions then arise. First of all, how easy is it really to find the sublists corresponding to those of TABLE 1.14 given the integer or symbol that labels the sublist (such as 2)?

Suppose that instead of comparing two symbols in the list we want to simply find the next element following or preceding a given element in a list. Does this sort of organization of the data help? How do we measure the degree of help provided by such a scheme? One very simple model of computation (direct access model) assumes that certain special classes of symbols are "directly accessible." Roughly speaking, this means that if such a symbol labels a sublist such as TABLE 1.14, then given the symbol, we can go directly to that sublist in a small constant amount of time independent of the "size of the problem" measured in some way. For example, imagine a very large version of TABLE 1.14. Given an integer n, we would assume in this model we can go directly to the list labeled by n in the same amount of time it took us to find the sublist labeled by 2 in the smaller TABLE 1.14. This is, of course, not true in reality. Just to read the digits of a *very* large number might take 1,000 years. A *very* large table such as TABLE 1.14 might be forced to have some of its entries at the bounds of the known universe. Nevertheless, experience shows that the basic idea of the direct access model is a good way to get a feeling for the complexity of certain basic computations. It will be useful at this point to look at one method for organizing computations with linearly ordered sets called *linked lists*.

Figure 1.15 shows an array of squares or "locations" in which we can store certain classes of basic symbols. The array is rectangular only for typographical convenience. Each square has an "address" such as A4, B6, etc. Given such an address, we can go to and read the contents of the designated square. In Figure 1.15, we are instructed to start at address D3 and follow the instructions contained in the square (as in the classical treasure hunt game). Doing this and reading the symbols we read @ # $ % ¢ &. Such a "data structure" is called a *linked list*.

1.15   LINKED LIST FOR LIST @ # $ % ¢ &.

Start at D3.

	1	2	3	4	5	6
A						$ GO TO B1
B	% GO TO D6				# GO TO A6	
C		&				
D			@ GO TO B5			¢ GO TO C2
E						

Figure 1.15

The basic assumption of the "direct access model" would be that, given a symbol such as "D5" or "E2," one can go to the corresponding region of the structure shown in 1.15 in "constant time." Courses in computational complexity can go into what this means in rigorous terms. For our purposes, the obvious intuitive ideas will suffice. The basic idea is that structures such as 1.15 can be allowed to grow in size over a fairly wide range with the same constant access time for all structures. This rough idea is all that is needed to get one started in thinking about "how efficient" certain computational ideas are. Studies of "computational complexity" can be overdone and carried to a point where they bear a rather questionable relationship to actual programming problems!

1.16   STRUCTURAL DIAGRAM OF FIGURE 1.15.

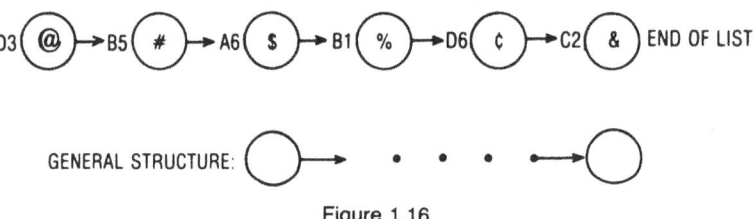

Figure 1.16

Associated with a "data structure" such as FIGURE 1.15 one can ask a number of basic questions. To get a more "global" or "geometric" feeling for the linked list of 1.15, one can represent the situation by a structural diagram such as shown in 1.16. The elements of the list are shown inside the circles; the "address" of each element is written just to the left of the circle. The arrows of 1.16 represent the fact that each location containing an element of the list also contains the address of the next entry in the list. The address of the next entry (represented by the arrow) is called a "pointer."

The reader should consider certain basic questions associated with the linked list data structure of 1.15. For instance, how would one carry out the instruction to "insert a new symbol * between $ and %?" Clearly, one would first locate an unused square in 1.15 (such as E6) and put the new symbol * in that square. One would then go to the square A6, which contains $, and replace the pointer "GO TO B1" with "GO TO E6." One would add to square E6 a pointer "GO TO B1." How much time is required to make these changes? First of all, we have to find an empty location or square. Do we have to scan a large number of squares to find one or can we be clever in how we keep track of empty squares so we can find one in "constant time?" Next, we have to find the square containing the symbol $. We have assumed that we can go to a square in constant

10

time given the *address* (such as "A6") but not given the *contents* "$." We can get around this problem if the symbols of the list are address symbols.

These problems that arise in our very simple-minded model can be magnified many times in dealing with a "real world" computational situation. If we are optimists, we can agree that the above computations required to insert a new symbol can be done in constant time. If we believe this in the context of the direct access model, then we would probably believe that the time required to insert a new symbol is *independent of the length of the list*. This follows, since one symbol is added and two pointers are changed independent of the length of the list. If our data structure consisted of typing each list symbol with one space between symbols, then we would represent the list of 1.15 by @ # $ % ¢ &. Now, inserting a new symbol * between $ and % would not, even viewed by an optimist, require constant time independent of the length of the list if the new list is to be represented in the same manner.

1.17   A MORE COMPLEX DATA STRUCTURE.

LIST 1: START E5   LIST 2: START D6   LIST 3: START D3   LIST OF LISTS: START E1

STRUCTURAL DIAGRAM
FOR LIST OF LISTS: E1 (D6) ⟶ D5 (E5) ⟶ E6 (D3) ⟶ E1 (START)

	1	2	3	4	5	6
A	D2 LIST 1	E2 LIST 1	E5 LIST 1 (START)	C3 LIST 1	D1 LIST 3	D3 LIST 3 (START)
B	D4 LIST 2	C1 LIST 2	C4 LIST 2	D6 LIST 2 (START)	C6 LIST 3	C2 LIST 3
C	B2 LAST ENTRY START D6	B6 AHEAD C6	A4 BACK D2 LAST ENTRY START E5	B3 AHEAD C1 START D6		B5 START D3
D	A5 AHEAD C2	A1 BACK E2 AHEAD C3	A6 AHEAD D1	B1 AHEAD C4 START D6	E5 (LIST 1) AHEAD E6	B4 AHEAD D4 FIRST ENTRY
E	D6 (LIST 2) AHEAD D5	A2 BACK E5 AHEAD D2			A3 FIRST ENTRY AHEAD E2 END C3	D3 (LIST 3) START E1

Figure 1.17   (Top)

11

STRUCTURAL DIAGRAMS:

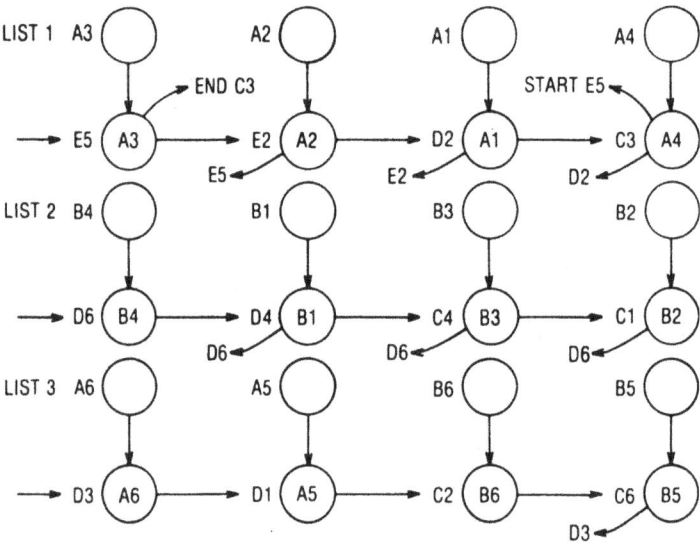

Figure 1.17   (Bottom)

A more complex data structure involving the idea of linked lists is shown in FIGURE 1.17. Three lists are stored in the 5 by 6 array shown. The symbols appearing in the lists are symbols such as "A2," "B3," etc. These symbols are also possible address symbols for squares in the array. We assume we can go to the addressed square in constant time for a wide range of such arrays. Thus, although the lists shown here are only of length 4, we can think of the analogous situation for lists of length n, n being much larger than 4. In Figure 1.17, LIST 1 is a "doubly linked" list in that each square containing a symbol of the list (except the first and last) has a pointer AHEAD to the location of the next symbol in the list and a pointer BACK to the previous symbol in the list. As the symbols stored in the list are also addresses, we have used those addresses to store (for each symbol in the list) its address *in the list*. Thus in A2 we find E2, which is the square that contains A2 in LIST 1. Also, the square containing the first entry in LIST 1 contains a pointer to the last entry in LIST 1 and vice versa. LIST 2 is a simple linked list where each square contains a pointer to the first square in the list. LIST 3 is a linked list with a pointer from the last entry to the first. FIGURE 1.17 also illustrates an important idea concerning how complex data structures may be manipulated using pointers.

In addition to the above mentioned lists, there is a list called LIST OF LISTS. The elements of this list are the three addresses to the first entries of LIST 1, LIST 2, and LIST 3. The LIST OF LISTS starts at E1 and contains the address

12

D6 of the first entry in LIST 2. Also in E1 we find a pointer AHEAD to the next square in LIST OF LISTS which is D5. This square contains the address E5 to the start of LIST 1 and a pointer AHEAD to E6. In E6 we find the last square in LIST OF LISTS, which contains the address D3 of the first entry of LIST 3, and a pointer back to the start of LIST OF LISTS.

The important thing to notice here is that the sequence of addresses D6, E5, and D3 defines a linear order on the lists: LIST 2, LIST 1, LIST 3. The reader should think carefully about this situation. How would a new list be added to the LIST OF LISTS? How would the order on the lists be changed? The basic advantage here is the ease with which these operations can be done in that only pointers to the lists are being moved and not the entries in the lists themselves. The reader should also think about what might happen here if the lists are allowed to get larger and more numerous. How does this model relate to one's own experience in programming? Is the direct access model really nonsense or does it give you some useful ideas? One thing to keep in mind in playing this game is that the amount of information stored in each square of an array such as that of FIGURE 1.17 *should be constant* (not depend on the number or length of lists). Also, the reader should try to develop a good notation for structural diagrams in the general case (arbitrary length list n in this example).

FIGURE 1.18 gives a suggestion for one possible notation. A good notation for structural diagrams of data structures is especially important as the data structures become more complex, as in the case of graphs.

1.18 STRUCTURAL DIAGRAMS FOR GENERAL CASE.

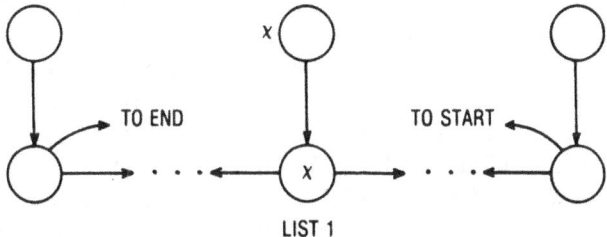

Figure 1.18

1.19 EXERCISE.

Discuss in detail how the data structure of FIGURE 1.17 would be modified in order to insert and delete elements from the various lists and change the order of the lists. Consider the general (m lists of length n) as m and n get large. Can these changes be made "in constant time?"

An important method for constructing linear orders on products of linear ordered sets involves the notion of lexicographic order. This is a familiar idea

13

to most readers but it can involve some subtle ideas when applied to combinatorial problems.

"Lexicographic order" derives its name from the order imposed on the words in a dictionary. As before, the set $\{1,2,\ldots,n\}$ will be denoted by $\underline{n}$. The *functions* from $\underline{d}$ to $\underline{r}$ will be denoted by $\underline{r}^{\underline{d}}$. We denoted such functions by length d strings of numbers from $\underline{r}$ ("one line" notation). For example, 1 3 1 is a function in $\underline{3}^{\underline{3}}$. This function sends $1 \to 1$, $2 \to 3$, and $3 \to 1$. Of course, 1 3 1 may be regarded as a function in $\underline{r}^{\underline{3}}$ for any $r \geq 3$. The *domain* of this function is the set $\underline{3}$. The *range* is $\underline{r}$. The *image* is the set $\{1,3\}$. The *coimage* is the *partition* $\{\{1,3\}, \{2\}\}$ of $\underline{3}$ (i.e., the collection of subsets of the domain on which the function takes on its various values). Two functions $i_1\ldots i_d$ and $j_1\ldots j_d$ are equal if $i_k = j_k$ for all k.

1.20 DEFINITION.

Given two functions $f = i_1\ldots i_d$ and $g = j_1\ldots j_d$, we scan from left to right until we find the first k such that $i_k \neq j_k$. If $i_k < j_k$, we say f is *lexicographically less* than g. If $i_k > j_k$, we say f is *lexicographically greater* than g. If $i_k = j_k$ for all k, then $f = g$.

1.21 REMARK.

The order defined by DEFINITION 1.20 is linear (why?). This order on $\underline{r}^{\underline{d}}$ is called the *lexicographic order*. We shall call it *lex order* for short. If we scan instead from right to left in DEFINITION 1.20, the resulting order will be called the *colex order*.

To find the next function in lex order after 3 3 1 3 1 2 3 in $\underline{3}^{\underline{7}}$ we increase by one the right-most value that can be increased and set all values further to the right to 1. Thus 3 3 1 3 1 3 1 is the next function. In colex order we increase the left-most value that can be increased and set all values further to the left to 1. Thus 1 1 2 3 1 2 3 is the next function in colex order. The functions in $\underline{3}^{\underline{3}}$ in lex order start off 1 1 1, 1 1 2, 1 1 3, 1 2 1, 1 2 2, 1 2 3, 1 3 1, etc. The functions in colex order start off 1 1 1, 2 1 1, 3 1 1, 1 2 1, 2 2 1, 3 2 1, 1 3 1, etc. The *permutations* (or *injections* or *one-to-one maps*) in lex order are 1 2 3, 1 3 2, 2 1 3, 2 3 1, 3 1 2, 3 2 1. In colex order they are 3 2 1, 2 3 1, 3 1 2, $\ldots$. More interesting are the *nondecreasing* functions. In lex order they come out 1 1 1, 1 1 2, 1 1 3, 1 2 2, 1 2 3, 1 3 3, 2 2 2, 2 2 3, 2 3 3, 3 3 3. In colex order they are 1 1 1, 1 1 2, 1 2 2, 2 2 2, 1 1 3, 1 2 3, 2 2 3, 1 3 3, 2 3 3, 3 3 3. *Notice that any subset of a linearly ordered set is automatically linearly ordered by the same order relation* (e.g., the *nondecreasing* functions as a subset of $\underline{3}^{\underline{3}}$).

1.22 EXERCISE.

(1) List all of the 27 functions in $\underline{3}^{\underline{3}}$ in lex and colex order. Find the nondecreasing functions in each list and think about the induced order on this subset. What

14

are the "mirror images" of these various lists (the lists obtained by writing each element in reverse order)?

(2) Write a computer program to list all permutations on $\underline{n}$ in lex order. Execute the program for n = 4, 5.

(3) List all of the *strictly increasing* functions in $\underline{5}^{\underline{3}}$ in lex order and colex order. These correspond to the subsets of size 3 from $\underline{5}$.

The notions of lex and colex order extend easily to a product of linearly ordered sets $A_1 \times \ldots \times A_n$. For example, to list in lex and colex order the elements of $\underline{3} \times \{a,b\} \times \{\alpha,\beta\}$, we start off $1\ a\ \alpha$, $1\ a\ \beta$, $1\ b\ \alpha$, $1\ b\ \beta$, etc. and $1\ a\ \alpha$, $2\ a\ \alpha$, $3\ a\ \alpha$, $1\ b\ \alpha$, etc. We assume the natural order on $\underline{3}$ and $a < b$, $\alpha < \beta$.

One can also define lexicographic order recursively. Consider the product $\mathscr{A}_n = A_1 \times \ldots \times A_n$ of linearly ordered sets A_i. If n = 1, then lex order on $\mathscr{A}_n$ is just the order on A_1. In general, $(a_1,\ldots,a_n)$ is lexicographically less than $(a_1',\ldots,a_n')$ if a_1 is less than a_1' in the order on A_1 *or* $a_1 = a_1'$ and $(a_2,\ldots,a_n)$ is lexicographically less than $(a_2',\ldots,a_n')$ in $A_2 \times \ldots \times A_n$.

1.23 EXERCISE.

(1) Define colexicographic order recursively. Prove that the recursive definitions of lex and colex order give the same linear orders as DEFINITION 1.20 and REMARK 1.21.

(2) Let $(L_1,\simeq)$ and $(L_2,\leq)$ be linearly ordered sets. Define a relation on $L_1 \times L_2$ by $(x,y)\ \underline{\alpha}\ (x',y')$ if $x \simeq x'$ in L_1 *and* $y \leq y'$ in L_2. By EXERCISE 1.13(3) this relation is an order relation. Consider $L_1 = \underline{2} \times \underline{2} = L_2$ with lexicographic order. Draw the Hasse diagram for the order on $L_1 \times L_2$.

The recursive definition of lex order gives a clear conceptual picture of how a card sorter works. Suppose we are given some subset $\mathscr{B}_n$ of $\mathscr{A}_n$. We want to put the elements of $\mathscr{B}_n$ into lexicographic order. To be a little more precise about the data structures involved, suppose that the elements (each a sequence of length n) of $\mathscr{B}_n$ are written on cards, one per card. Suppose we have $A_1 = \ldots = A_n = \underline{m}$ to simplify the notation. Imagine m buckets into which the cards can be placed. Given the cards of $\mathscr{B}_n$ in some order, they will be distributed into the various buckets according to some rules to be explained below. The critical idea will be that cards are put into the buckets *carefully* so that within each bucket the cards are stacked in such a way as to preserve the *same relative order* that they just had. The sorting algorithm is to first place the cards into buckets according to their right-most symbols. The cards are then removed from the buckets and stacked together (concatenated), cards from bucket 1 first, then bucket 2, etc. This process is repeated on the next symbol to the left, then the next, etc. The process is illustrated in FIGURE 1.24 for n = 2.

15

1.24 LEXICOGRAPHIC BUCKET SORT.

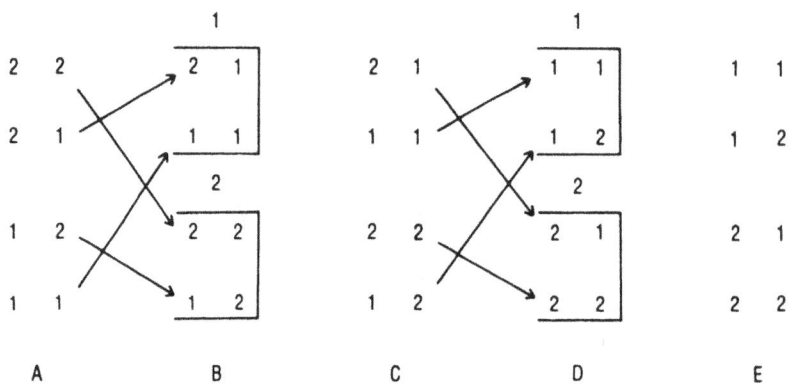

Figure 1.24

A. Reading from top to bottom, the list is in some order but not lex order.

B. The elements of the list are placed into "buckets" according to the right-most value. In each bucket the order is the same as the starting order. Thus in bucket 1 the element 2 1 is before 1 1 as in the starting order.

C. The contents of the buckets are concatenated.

D. The elements are placed in buckets according to the value of next symbol to the left. The order in C is preserved within each bucket.

E. The buckets are again concatenated and the list is now in lex order.

1.25 EXERCISE.

Give a careful proof that the lexicographic bucket sort algorithm will correctly put any sublist $\mathcal{B}_n$ of $\underline{m} \times \ldots \times \underline{m}$ into lex order.

Next we shall consider some variations on the idea of lexicographic order and related combinatorial problems.

Consider $\underline{r}^{\underline{d}}$ again. Introduce a new symbol $*$, which we add to the set $\underline{r}$. First assume $*$ precedes all elements of $\underline{r}$ so $\underline{r} \cup \{*\} = \bar{\underline{r}}$ is linearly ordered. Consider the usual lexicographic order on $\bar{\underline{r}}^{\underline{d}}$. We are interested in the subset $W_{d,r} \subseteq \bar{\underline{r}}^{\underline{d}}$ of functions of the form $i_1 i_2 \ldots i_k ** \ldots *$, $k = 1,2, \ldots, d$, where $i_t \in \underline{r}$ for $t = 1, \ldots, k$. In these strings (or functions) $*$ acts as a "terminal symbol" in the sense that if it appears at all in a position, then all subsequent positions contain $*$. For example, look at $\underline{2}^{\underline{3}}$. Here $\underline{2} = \{*, 1, 2\}$. In lex order $W_{3,2}$ becomes $1 * *$, $1 1 *, 1 1 1, 1 1 2, 1 2 *, 1 2 1, 1 2 2, 2 * *, 2 1 *, 2 1 1, 2 1 2, 2 2 *, 2 2 1$, $2 2 2$. A common way to graphically represent $W_{3,2}$ is shown in FIGURE 1.26(a) (the terminal $*$'s are omitted). The order in which $W_{3,2}$ appears in the above lex order is shown in FIGURE 1.26(b).

16

1.26 TREE DIAGRAM OF PRELEX ORDER.

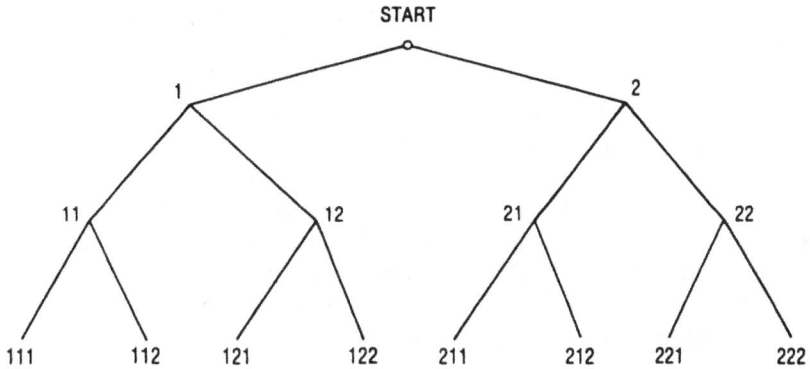

Figure 1.26a

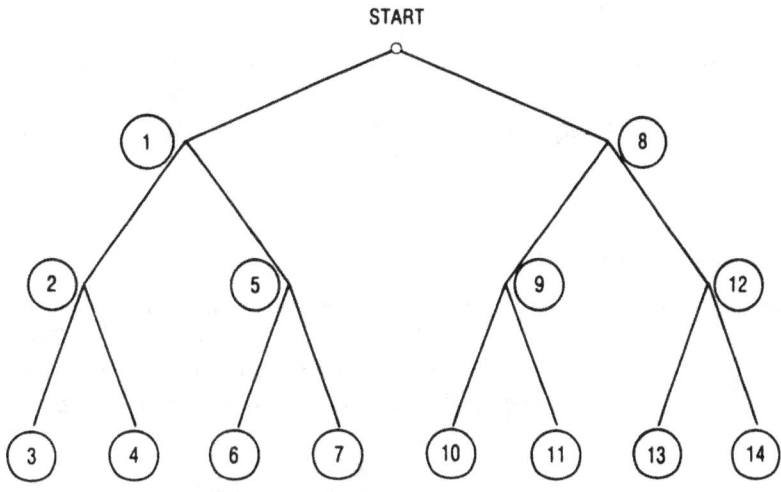

Figure 1.26b

1.27 EXERCISE.

Draw the analogous figures to 1.26(a) and 1.26(b) for the cases where ∗ is located {1,∗,2} and {1,2,∗} in the order on $\underline{2}$.

In general, let $W_{d,r}$ (words of length at most d on r symbols) denote the elements of $\bar{r}^{\underline{d}}$ where $\bar{r} = \underline{r} \cup \{*\}$ and $*$ is a terminal symbol. There are $r+1$ ways that we might insert $*$ into the linearly ordered set $\underline{r}$. Each of these gives rise to a lexicographic order on $W_{d,r}$. If $*$ is placed before all the symbols in $\underline{r}$, we call the order on $W_{d,r}$ the *prelex order*. If it is placed after, we call the order the *postlex order*. The rest of the orders we could call "inlex orders." These terms are in keeping with similar notions in computer science (see Knuth, D.E., *The Art of Computer Programming*, Vol. 1, Reading, MA: Addison-Wesley, p. 316). Diagrams such as FIGURE 1.26(a) and (b) are called "tree diagrams." In case $r = 2$ the tree is called "binary." There is only one inlex order in this case. The orders prelex, inlex, and postlex are commonly called "preorder," "inorder," and "postorder" in this case. These orders are illustrated in FIGURE 1.26(a) and (b), and in EXERCISE 1.27.

1.28 EXERCISE.

Draw enough of the analogs for FIGURE 1.26(a) and (b) for $W_{3,3}$ for prelex $*$ 1 2 3, postlex 1 2 3 $*$, and the two inlex orders 1 $*$ 2 3 and 1 2 $*$ 3 to get a feel for what is happening.

Another linear order that arises is length-first lex (or colex) order. If $p < q$ then any string $i_1 i_2 \ldots i_p ** \ldots *$ (of "length" p) is less than any string $i_1 i_2 \ldots i_q * \ldots *$ of length q. Among strings of equal length we use the lex (or colex) order.

1.29 EXERCISE.

The nonempty subsets of $\underline{n}$ correspond to the strictly increasing strings of $W_{n,n}$. For n = 4, list these strings in prelex and in length first lex order. The string $* * \ldots *$ corresponds to the empty set and precedes all elements of $W_{n,n}$ in both orders. ("strictly increasing" here applies only to numerical symbols, not $*$'s).

The above linear orders will arise later (in a slightly different form) in our discussion of graphs and trees. Postlex order will be called *postorder*, prelex order will be called *preorder*, and length-first order will be called *breadth-first order*.

The various lexicographic orders on $W_{d,r}$ are inherited by any subset $S \subseteq W_{d,r}$. Also, the discussion of these orders applies if $\underline{d}$ or $\underline{r}$ is replaced by any finite linearly ordered set. The discussion of prelex (or postlex) order applies equally well to a direct product of arbitrary linearly ordered sets $A_1 \times \ldots \times A_d$ as the symbol $*$ may be addended to the beginning (or end) of all of the sets A_i. In particular, the linear order on A_i may itself be a lex, colex, prelex, etc., order.

1.30 EXAMPLE.

In FIGURE 1.31(b) we have indicated the list of all possible positions for five different "tiles" that might be placed on the 4×4 board shown in FIGURE 1.31(a). The numbers in FIGURE 1.31(b) indicate the squares of FIGURE 1.31(a) covered by that position of the tile (read as in a book).

1.31 LEX LISTS OF TILINGS OF A BOARD (backtracking problem).

(a)

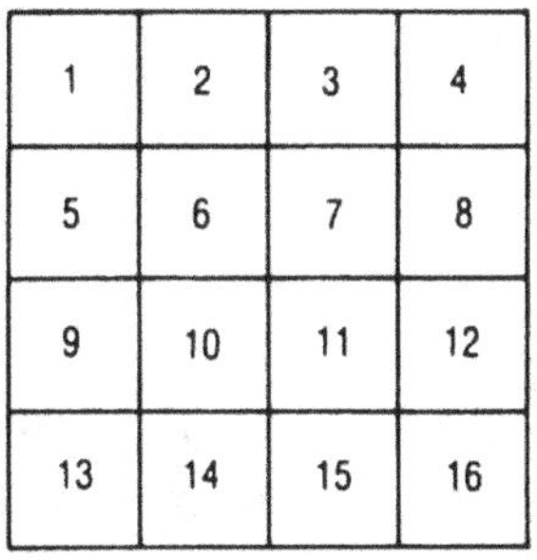

1	2	3	4
5	6	7	8
9	10	11	12
13	14	15	16

$\tilde{A}_i = \{*\} \cup A_i$,

A SEQUENCE SUCH AS
$((1,2,3,6), (5,9,10,14), *,*,*)$
IS WRITTEN
$((1,2,3,6),(5,9,10,14))$ AND IS
SAID TO HAVE
"LENGTH 2."

(b)

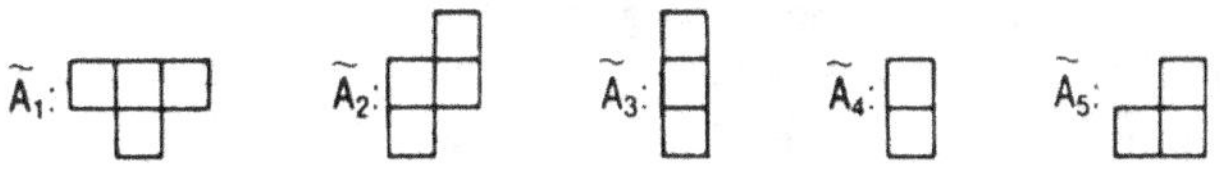

$\tilde{A}_1$:	$\tilde{A}_2$:	$\tilde{A}_3$:	$\tilde{A}_4$:	$\tilde{A}_5$:
*	*	*	*	*
(1,2,3,6)	(1,2,6,7)	(1,2,3)	(1,2)	(1,2,5)
(1,5,6,9)	(1,5,6,10)	(1,5,9)	(1,5)	(1,2,6)
(2,3,4,7)	(2,3,5,6)	(2,3,4)	(2,3)	(1,5,6)
(2,5,6,7)	(2,3,7,8)	(2,6,10)	(2,6)	(2,3,6)
(2,5,6,10)	(2,5,6,9)	(3,7,11)	(3,4)	(2,3,7)
$\vdots$	$\vdots$	$\vdots$	$\vdots$	$\vdots$

Figure 1.31

Observe that each list A_1 is ordered lexicographically. The natural order on $A_1 \times \ldots \times A_5$ is the prelex order ($\underline{\mathbb{r}}^d$ replaced by $\tilde{A}_1 \times \ldots \times \tilde{A}_5$). In covering the board with these "tiles," only strings with all sets disjoint—such as the length 2 string, $(1,2,3,6)$, $(5,9,10,14)$ (FIGURE 1.32(c)), or length 5 string, $(1,2,3,6)$ $(4,7,8,11)$, $(5,9,13)$, $(10,14)$, $(12,15,16)$ (FIGURE 1.32(a))— will even be considered. In prelex order, the string $(1,2,3,6)$, $(5,9,10,14)$ would be encountered before any longer string starting off with the same two terms. This is desirable because we see (FIGURE 1.32(c)) that this string leaves the square number 13 isolated. Hence no continuation of this string can lead to a solution. Thus we immediately take a *leap forward* in the prelex order on $A_1 \times \ldots \times A_5$, skipping many strings. This technique of testing and leaping forward is called (strange as it may seem) "backtracking."

19

1.32 SAMPLE TILING CONFIGURATIONS.

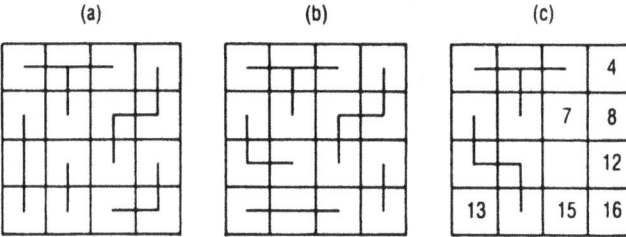

Figure 1.32

1.33 EXERCISE.

(1) Add the next five entries to each of the sets A_i, $i = 1, \ldots, 5$, in FIGURE 1.31(b). In FIGURE 1.32, is (a) or (b) smaller in lex order? Is there any solution to this tiling problem smaller than both (a) and (b)?

(2) The 3×3 board below is to be covered with the tiles shown below just as in FIGURE 1.31. Make up the lexicographic lists A_1, A_2, and A_3 analogous to FIGURE 1.31. List all elements of $A_1 \times A_2 \times A_3$ that are coverings of the board in lexicographic order.

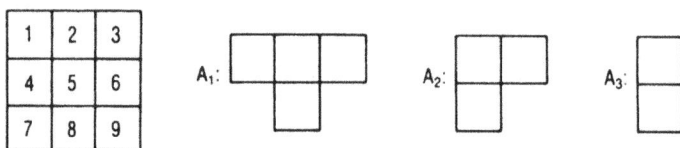

Figure 1.33   (Top)

(3) Solutions to the problem of placing five nonattacking queens on a 5×5 chessboard are specified by sequences such as $(1,4,2,5,3)$ and $(4,1,3,5,2)$. The numbers represent the location of the queen in the respective columns of the board:

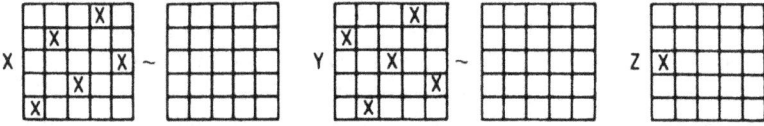

Figure 1.33   (Bottom)

(a) Two solutions are equivalent if one can be made to coincide with the other by rotating and/or reflecting the board. Are X and Y above in the same equivalence class? Explain. If a lexicographically minimal system of representatives is to be selected, what element or elements would be selected equivalent to X and Y? (See EXERCISE 1.38(2) for related ideas.)

(b) Show that there is no minimal representative that starts off (3,. . .) as in Z above.

1.34 EXERCISE.

The "Mathematical Games" section of *Scientific American* (Vol. 227, No. 3, September 1972, pp. 176–182) describes a game called SOMA cube (Parker Brothers, Inc., Salem, Massachusetts). How would you linearly order the solutions and partial solutions to this game?

1.35 EXAMPLE.

The string of symbols v v h h h v v h describes a covering of the 4 × 4 board of FIGURE 1.36(a) with dominoes (or "dimers"). To interpret the string, read the first symbol, a v, and place a vertical domino on the square numbered 1. The second symbol is a v so place a vertical domino on the lowest numbered uncovered square (number 2). The third symbol is an h so place a horizontal domino on the lowest uncovered square, etc. The configuration after the first four symbols of the string have been interpreted is shown in FIGURE 1.36(b) and the final configuration in FIGURE 1.36(c). Of course, not all of the 2^8 = 256 strings of length 8 represent coverings of the square with dominoes (v v v h . . . wouldn't!) There are 36 "grammatically correct" strings that do represent coverings.

If we let A = {h,v} be a two-point ordered set with h < v, then the strings of length 8 or less (call them $W_{8,A}$) are ordered by the prelex order. In particular, the solutions (all of length 8) are ordered by the lex order on A^8. The 36 solutions are shown in FIGURE 1.37. They are not in lex order.

1.36 CODING DOMINO COVERINGS.

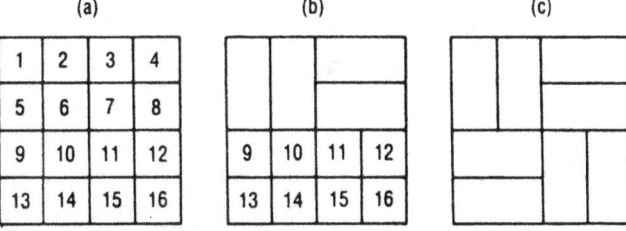

Figure 1.36

21

1.37 ALL DOMINO COVERINGS OF A 4 × 4 BOARD.

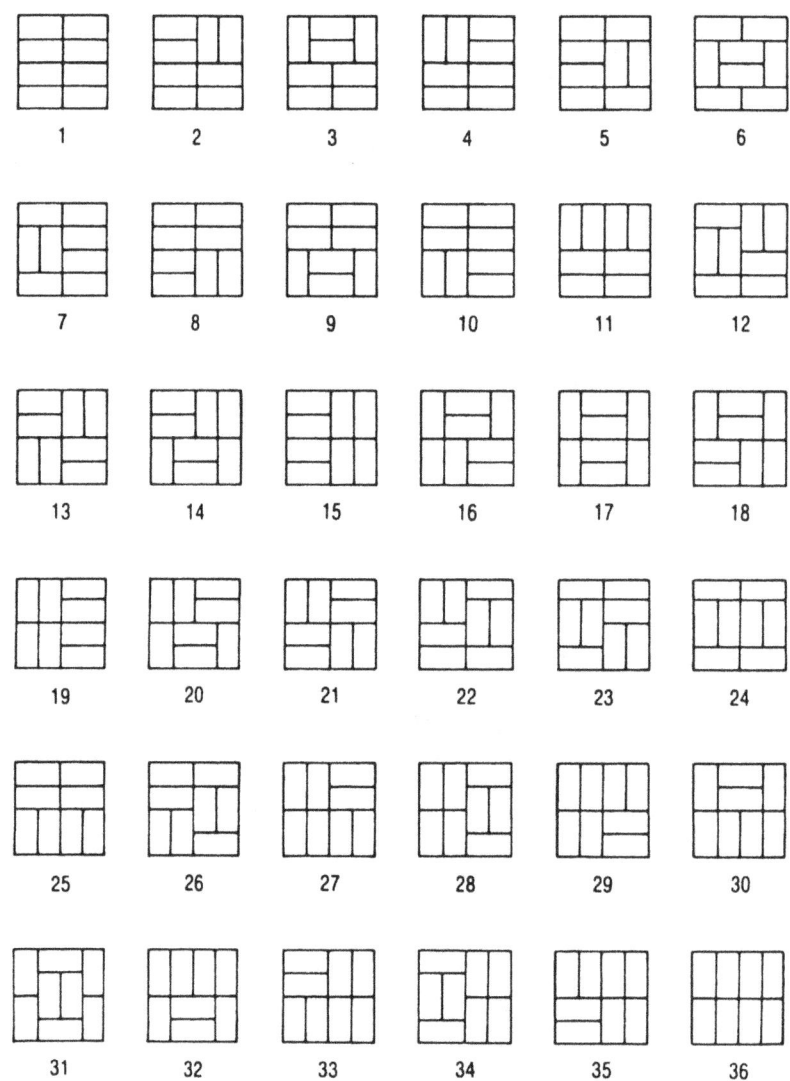

Figure 1.37

1.38 EXERCISE.

(1) Find the next to last solution of FIGURE 1.37 in lex order. Carry out a systematic procedure for putting the solutions of FIGURE 1.37 in correct lex order.

(2) Look at solution 27 of FIGURE 1.37. A rotation by 90° produces a lex smaller figure, so cross out 27. Look at solution 25. It is the smallest among all possible rotations (90°, 180°, 270°) and reflections of itself. So keep 25. Do this for all of the 36 figures. Verify that the final list of solutions that you keep by this procedure has the following properties (such lists are called "systems of representatives"):

    (a) No two solutions are "isomorphic" in the sense that a rotation and/or reflection of one produces the other.

    (b) All of the 36 solutions are "isomorphic" to some solution in your list.

       Problems of selecting sublists satisfying (a) and (b) with respect to groups of symmetrics are called *isomorph rejection* problems. We will discuss them later.

(3) Think up a totally different algorithm than EXERCISE 1.38(2) to perform isomorph rejection in FIGURE 1.37. Which algorithm is easier for a standard computer? Which is easier for you?

(4) Invent your own algorithms for making a list of all possible ways to color the faces of a tetrahedron and a cube with r different colors up to (a) rotations and (b) rotations and reflections. First try r = 2, 4, 6.

(5) List all ways of coloring the vertices of a hexagon with three colors, B, G, R, up to rotations and reflections, such that adjacent vertices are never given the same color. How many such figures are there for r colors?

(6) Try to write some programs along the lines of the previous problems.

Conceptually, there is another way to regard the set $W_{d,r}$. Given a string $i_1 i_2 \ldots i_t$ we can think of it as a key or label for a *subset* of $\underline{r}^d$. The subset is the set of all $j_1 j_2 \ldots j_d$ with $j_1 \ldots j_t = i_1 \ldots i_t$. Consider $W_{3,2}$. The strings 1 and 2 are associated with the sets $B_1 = \{1\ 1\ 1,\ 1\ 1\ 2,\ 1\ 2\ 1,\ 1\ 2\ 2\}$ and $B_2 = \{2\ 1\ 1,\ 2\ 1\ 2,\ 2\ 2\ 1,\ 2\ 2\ 2\}$. The strings 1 1, 1 2 are associated with the sets $B_{11} = \{1\ 1\ 1,\ 1\ 1\ 2\}$ and $B_{12} = \{1\ 2\ 1,\ 1\ 2\ 2\}$. The strings 21 and 22 are associated with the sets $B_{21} = \{2\ 1\ 1,\ 2\ 1\ 2\}$ and $B_{22} = \{2\ 2\ 1,\ 2\ 2\ 2\}$. The string 1 1 1 is associated with the one element set $B_{111} = \{1\ 1\ 1\}$ consisting of itself, as are the rest of the length 3 strings. An *ordered partition* of a set S is a sequence $(Q_1, \ldots, Q_p)$, $p \geq 1$, of subsets (possibly empty) of S, pairwise disjoint, with union S. Thus (B_1, B_2) is an ordered partition of $\underline{r}^d$ and (B_{11}, B_{12}) is an ordered partition of B_1, etc. The ordered partition $(B_{11}, B_{12}, B_{21}, B_{22})$ is an ordered partition of $\underline{r}^d$ which is a *refinement* of (B_1, B_2) in the sense that each of its subsets is contained in some subset of the ordered partition (B_1, B_2). This conceptual association between strings and successive refinements by ordered partitions is quite common in mathematics. For example the binary decimal .1 0 1 refers to the set $B_1 = \{x: 1/2 \leq x < 1\}$, refined by $B_{10} = \{x: 1/2 \leq x < 3/4\}$, refined by $B_{101} = \{x: 5/8 \leq x < 3/4\}$, refined by $\{x: 5/8 \leq x < 11/16\}$, etc. The number of refinements corresponding to .10100 . . . is infinite, defining in the limit one real number. There is a conceptually natural way to associate linear orders with successive refinements by ordered partitions. We now explore this method.

FIGURE 1.39 shows a structure that we will call an *ordered partition tree* for the set S (in this case S = $\underline{6}$). We shall give a more formal definition of the idea of an *ordered tree* later. As for now, we shall use this idea intuitively to discuss linear orders on sets.

1.39 ORDERED PARTITION TREE.

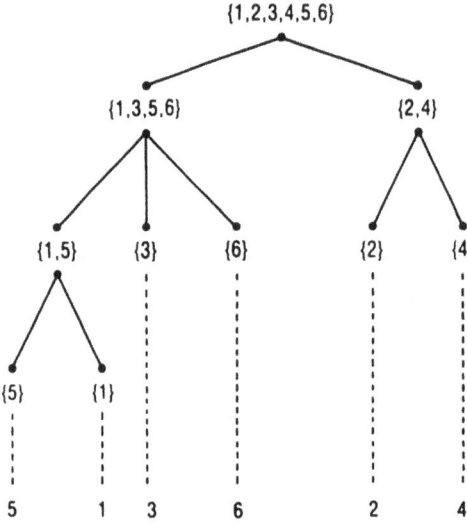

Figure 1.39

The *vertices* of the tree of FIGURE 1.39 are nonempty subsets of $\underline{6}$. The set $\underline{6}$ is the *root* of the tree (the tree is upside down). The root has two *sons*: {1, 3, 5, 6} (the first son) and {2, 4} the second son. The sons of each vertex define an ordered partition of that vertex. None of the sets in the ordered partition are empty. The vertices such as {5},{1},{3}, etc., are *terminal vertices* or *leaves* in the tree. Each terminal vertex corresponds to a subset of $\underline{6}$ with one element. The dotted lines project the terminal vertices onto a horizontal line. When reading the elements of $\underline{6}$ as they appear on the horizontal line one obtains 5, 1, 3, 6, 2, 4. This defines the linear order associated with the ordered partition tree of FIGURE 1.39.

A given order, such as 5, 1, 3, 6, 2, 4, may be defined by many different ordered partition trees. The *nodes* or *vertices* in FIGURE 1.39 are labeled by subsets of $\underline{6}$. Note that only the labels of the leaves need be given to recover all of the other labels. That is, each label of an internal node is exactly the set of elements in the leaves descendant to that label. There is one leaf corresponding to each element in the set associated with the root of the partition tree ($\underline{6}$ in FIGURE 1.39). Why bother to label the internal nodes at all if they are deter-

24

mined by the leaves? The reason is that we may wish to define the partition tree "locally" by giving rules for constructing the tree rather than by drawing the whole tree, as in FIGURE 1.39. We now discuss some examples of this type of situation.

Consider $\underline{r}^{\underline{d}}$ as above. Let $(i_1, \ldots, i_t, \_, \ldots, \_)$ denote the set of all functions in $\underline{r}^{\underline{d}}$ whose first t values are the ones specified. In our above notation, $B_1 = (1, \_, \_)$. Note that in a tree diagram such as FIGURE 1.39 there is a unique path (sequence of edges) from the root to any given node of the tree. We call the number of edges in that path the *height* of that node. Figure 1.40 gives a local description of one possible partition tree for $\underline{r}^{\underline{d}}$.

1.40   LOCAL DESCRIPTION OF A PARTITION TREE FOR $\underline{r}^{\underline{d}}$.

(1) ROOT $= (\_, \ldots, \_) = \underline{r}^{\underline{d}}$
(2) INTERNAL NODES:

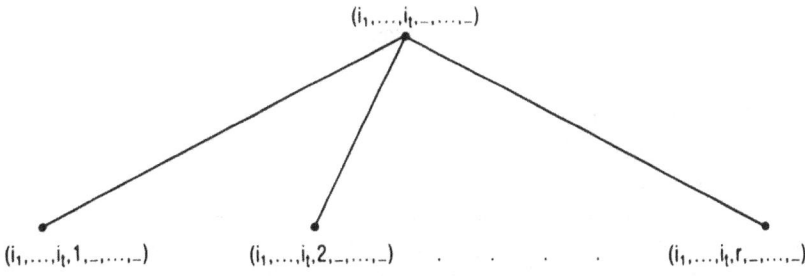

LEXICOGRAPHIC ORDER

Figure 1.40

(3) ANY NODE OF HEIGHT d IS A LEAF.

The order defined by the partition tree of FIGURE 1.40 is lexicographic order on $\underline{r}^{\underline{d}}$. The case r = 2 and d = 3 is shown in FIGURE 1.41.

1.41   PARTITION TREE FOR $\underline{2}^{\underline{3}}$.

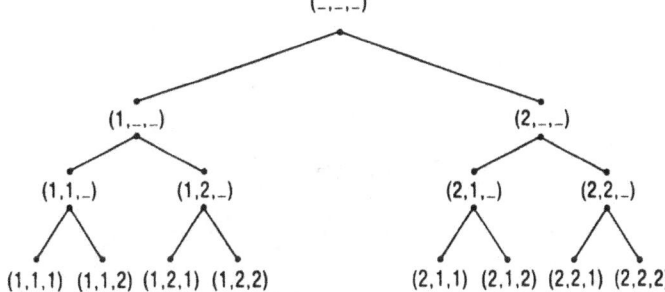

Figure 1.41

25

A tree is a *binary tree* if each internal node has one or two sons, a *full binary tree* if each internal node has exactly two sons, and a *complete binary* tree if it is full and all leaves are at the same level (see FIGURE 1.42 for an example).

1.42 LEXICOGRAPHIC ORDER ON $\underline{2}^3$.

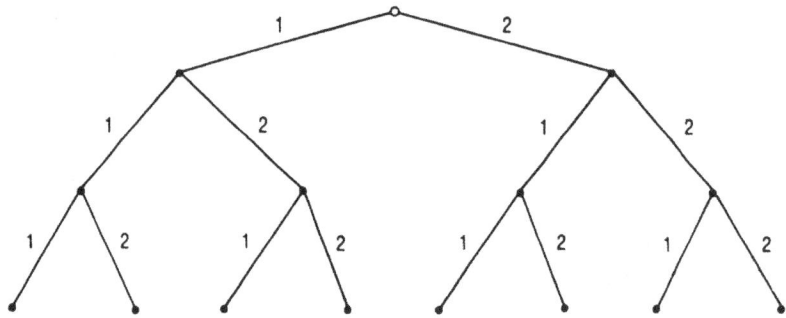

Figure 1.42

The complete binary tree of FIGURE 1.42 is just an obvious modification of that of FIGURE 1.41. Start at the root of the tree in FIGURE 1.42 and follow a path to a leaf, reading the labels of the edges traversed. Each leaf has height 3 and thus is assigned a triple (i,j,k). Reading the associated triple of the leaves as they occur from left to right gives $\underline{2}^3$ in lexicographic order.

Observe that in FIGURE 1.42 the left and right sons of the root are themselves roots of complete binary trees of height 2. These two subtrees are identical, so they may be superimposed on each other as seen in FIGURE 1.43. In this figure, paths from the root to the terminal nodes still correspond to elements of $\underline{2}^3$ in lex order. The resulting "graph" is no longer a tree.

1.43 A MODIFICATION OF FIGURE 1.42.

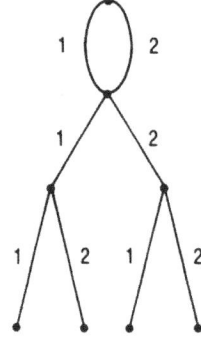

Figure 1.43

The two subtrees of FIGURE 1.43 may be superimposed in the same manner and the terminal vertices identified to get FIGURE 1.44. Paths from the topmost vertex in this figure to the bottom vertex still correspond to paths in $\underline{2}^3$. To describe the linear order (lex order) in terms of FIGURE 1.44 one must essentially give the definition of lex order (DEFINITION 1.20).

1.44   A MODIFICATION OF FIGURE 1.43.

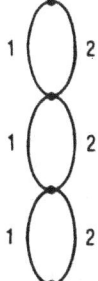

Figure 1.44

We shall now consider partition trees for some additional linear orders. Let S_n denote the set of all permutations from $\underline{n}$ to $\underline{n}$. Using notation similar to that above, let $(i_1, \ldots, i_k, \_, \ldots, \_)$ denote the set of all permutations in S_n with first k values as specified. We let $S_n = (\_, \ldots, \_)$. The partition tree of FIGURE 1.45 is associated with lex order on S_3.

1.45   LEX ORDER ON S_3.

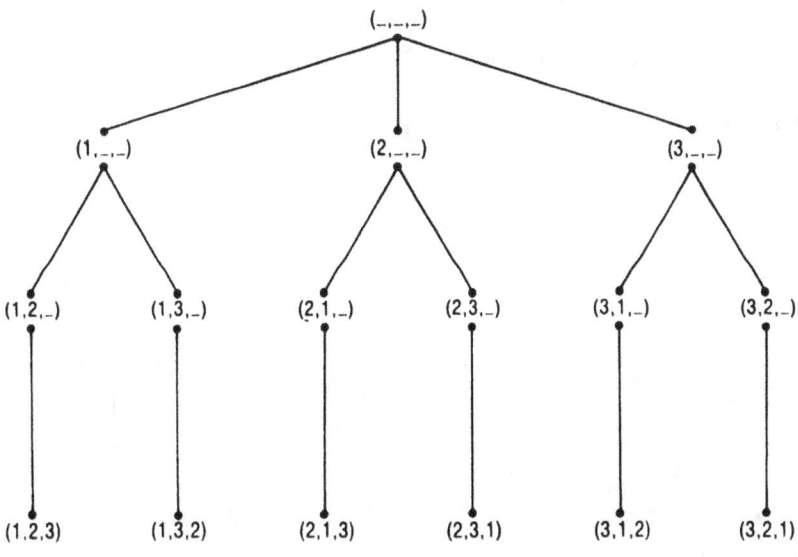

Figure 1.45

27

1.46 EXERCISE.

(1) Consider the general case of a partition tree for S_n as suggested by the example for S_3 in FIGURE 1.45. Give a careful local description of the general case along the lines of FIGURE 1.40.

(2) What is the analog of FIGURE 1.44 for FIGURE 1.45 and for the general case? *Hint*: Note that all subtrees at the same level are identical.

Another basic partition tree for S_n is based on the "direct insertion" method for generating permutations. Consider the permutation 426153 in S_6. Note that the 2 is to the left of the 1. We say that "2 is in position two relative to 1." If the 2 were to the right of 1 we would say "2 is in position one relative to 1." For any permutation in S_n we consider the two possible positions of the 2 relative to the 1: $\frac{-}{2} 1 \frac{-}{1}$. The position on the right is called the first position, the other position is the second position. Let $(2 \text{ in } 1)_n$ denote the subset of S_n of all permutations with 2 in position 1 and let $(2 \text{ in } 2)_n$ denote all permutations with 2 in position 2. Thus, $465213 \in (2 \text{ in } 2)_6$ and $4123 \in (2 \text{ in } 1)_4$. Similarly, if we consider a permutation in S_n, $n \geq 3$, then we define 3 positions of 3 relative to the symbols 1 and 2: $\frac{-}{3} x \frac{-}{2} y \frac{-}{1}$. The x and y can be either 1 or 2. Thus 4231 and 4132 are both elements of the set $(3 \text{ in } 2)_4$, 4321 and 4312 are both in $(3 \text{ in } 3)_4$, etc. In the same manner, we define sets $(k \text{ in } j)_n$, $j \in \underline{k}$. Another way to view the set $(k \text{ in } j)_n$ is as follows: Given a permutation in S_n, and an integer k, remove all symbols greater than k from the permutation. The permutation is in $(k \text{ in } j)_n$ if k is the jth symbol from the right in the reduced permutation. Consider 957862431 and $k = 7$. The reduced permutation is 5762431 and 7 is the 6th symbol from the right. Thus, $957862431 \in (7 \text{ in } 6)_9$. A partition tree for S_3 based on this idea is shown in FIGURE 1.47.

1.47 DIRECT INSERTION PARTITION TREE FOR S_3.

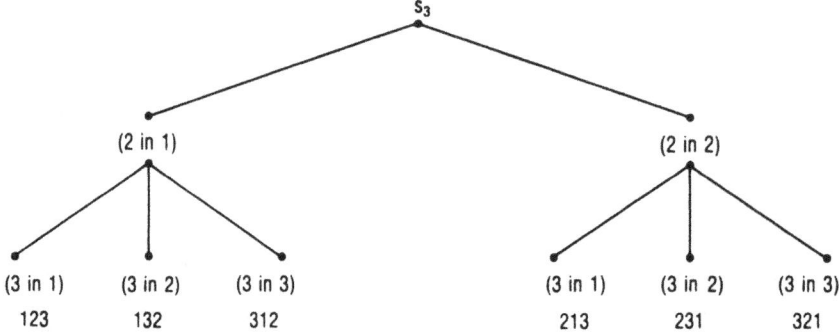

Figure 1.47

28

The labels in the partition tree of FIGURE 1.47 are not the actual partitions (except labels of sons of the root). In FIGURE 1.47 and in its generalization to S_n, the partition associated with a vertex v is obtained by taking the intersection of all sets labeling vertices on the path from the root to v. The intersection of all sets on a path from the root to a leaf is a set consisting of a single permutation. Thus the order of the leaves from left to right defines a linear order on S_n. For example, (2 in 1) $\cap$ (3 in 3) = {312} in FIGURE 1.47. The permutations associated with each leaf are shown in FIGURE 1.47.

Note that in the first three permutations of FIGURE 1.47, the "1,2" pattern remains fixed and the 3 is inserted in the three positions right to left relative to this pattern. Then, the pattern becomes "2,1" and the 3 is again inserted right to left. This basic idea extends to the case of S_n and is the reason for the name "direct insertion method." Another way of representing the tree in FIGURE 1.47 is shown in FIGURE 1.48.

1.48   CODED VERSION OF DIRECT INSERTION TREE.

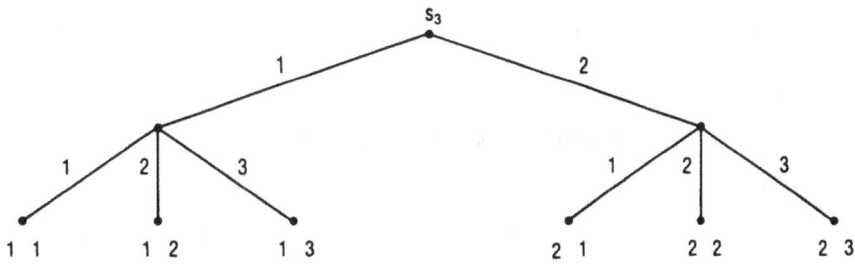

Figure 1.48

In FIGURE 1.48, the labels on the edges tell where the corresponding integer is to be inserted. In any path from the root to a leaf, the first edge tells where to insert the symbol 2, the second edge where to insert 3, etc. This tree represents the same linear order on permutations as the direct insertion tree. It is interesting to label each leaf of the tree of FIGURE 1.48 with the sequence of edge labels obtained by following the path from the root to that leaf. This has been done in FIGURE 1.48. Note that the labels on the leaves, read left to right, is just lex order on $\underline{2}$ x $\underline{3}$. Thus there is a natural bijection between S_n in direct insertion order and $\underline{2}$ x . . . x $\underline{n}$ in lex order. This is an example of an *order isomorphism*. Between linearly ordered sets, only one bijection preserves order.

1.49   EXERCISE.

(1) Extend FIGURES 1.47 and 1.48 to S_n by giving the local description such as in FIGURE 1.40 or EXERCISE 1.46(1).

(2) Give a careful proof of the correspondence between S_n in direct insertion order and $\underline{2} \times \ldots \times \underline{n}$ in lex order.

(3) What is the successor and predecessor of 87612543 in lex and direct insertion order on S_8? What is the first entry in the second half of the list of S_8 in lex and direct insertion order?

As one can see from the above examples, the idea of "partition tree" has many different variations. The general idea is that given a set S and a linear order on S, we wish to construct some geometric object that aids our intuition. Thus, we seek a tree or graph together with a bijection between certain classes of paths and the elements of our set S. The idea of a partition tree is a general idea that can be modified as needed in various particular situations. We have discussed permutations above. Now we consider "combinations" or subsets of fixed size k from a set of size n. Let $\mathcal{P}_k(n)$ denote the set of all subsets of size k chosen from $\underline{n}$. Let $D(\underline{n}^k)$ denote the set of all decreasing functions from $\underline{k}$ to $\underline{n}$. For example, $D(\underline{4}^{\underline{3}}) = \{321,421,431,432\}$ where the functions are listed in one line notation and in lexicographic order. Clearly, to list the elements of $\mathcal{P}_k(\underline{n})$ it suffices to list $D(\underline{n}^k)$ as there is a natural bijection between the two sets (for example, the decreasing function 431 corresponds to the set $\{4,3,1\}$).

1.50 EXERCISE.

For $f \in D(\underline{n}^k)$ define $F(f) = \text{Image}(f)$. Give a rigorous proof that F is a bijection between $D(\underline{n}^k)$ and $\mathcal{P}_k(\underline{n})$.

Observe that $D(\underline{n}^k)$ is the disjoint union of $D(\underline{n-1}^k)$ and the set of all decreasing functions in $D(\underline{n}^k)$ that begin with n. We denote this latter set by $nD(\underline{n-1}^{k-1})$ as all such functions in one line notation consist of n followed by a decreasing function from $\underline{k-1}$ to $\underline{n-1}$. In the same manner, $D(\underline{n-1}^k)$ is a disjoint union of $D(\underline{n-2}^k)$ and $(n-1)D(\underline{n-2})^{k-1}$. Repeating this process we arrive at a partition tree with local description as shown in FIGURE 1.51.

1.51 PARTITION TREE FOR DECREASING FUNCTIONS.

(1) Root $D(\underline{n}^k)$

(2) INTERNAL NODES

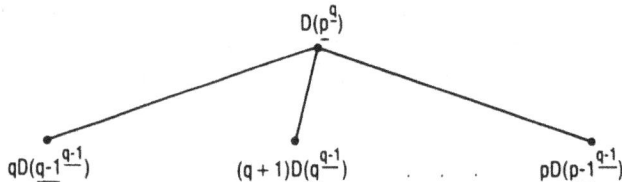

(3) ANY NODE OF HEIGHT k IS A LEAF

Figure 1.51

30

The basic idea of FIGURE 1.51 is very simple, the decreasing functions from $\underline{k}$ to $\underline{n}$ are decomposed first into all functions that start with k, then all functions that start with k + 1, etc. It is better to represent this process in a manner analogous either to FIGURE 1.42 or 1.48. This is done in FIGURE 1.52.

1.52 TREE DIAGRAM FOR D($\underline{6}^4$) IN LEX ORDER.

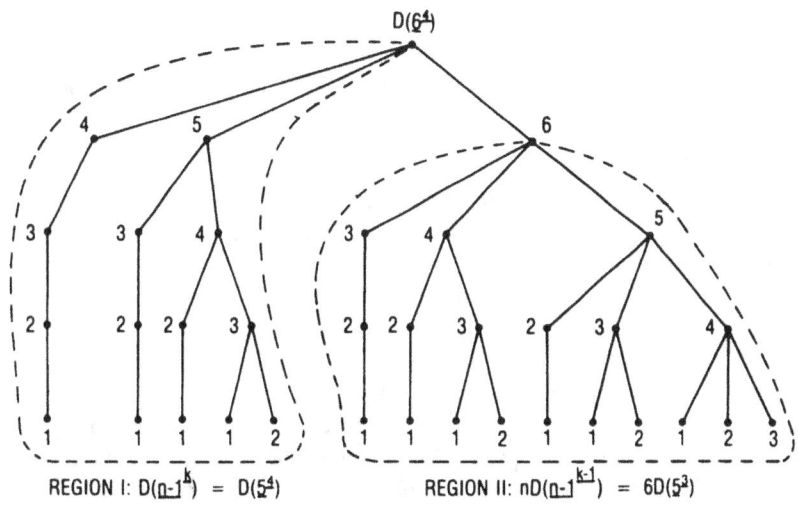

Figure 1.52

The interpretation of FIGURE 1.52 is the obvious one—start at the root and descend to a leaf, reading the labels on the vertices encountered on the way. Thus, each terminal node corresponds to a unique decreasing function (the sequence of labels from the root to the terminal node). The order defined by left to right order on terminal nodes is lex order on decreasing functions from $\underline{4}$ to $\underline{6}$. The sons of the root are labeled 4, 5, and 6 and the corresponding subtrees rooted at these vertices give all decreasing functions of length 4 starting with 4, 5, and 6, respectively (as specified in FIGURE 1.51). The basic recursion that D($\underline{n}^k$) is the disjoint union of D($\underline{n-1}^k$) and nD($\underline{n-1}^{k-1}$) is shown by the dotted REGION I and REGION II of FIGURE 1.52.

1.53 EXERCISE.

(1) Try to prove the standard result $|\mathscr{P}_k(\underline{n})| = \dfrac{n!}{k!(n-k)!}$ without consulting your previous notes or textbook.

(2) FIGURES 1.51 and 1.52 show geometrically that $\binom{n}{k} = \sum_{j=k-1}^{n-1} \binom{j}{k-1}$ and $\binom{n}{k} = \binom{n-1}{k-1} + \binom{n-1}{k}$. Prove these results by induction using the formula for $\binom{n}{k}$ in problem (1) above (assume $k \geq 1$).

(3) Consider the process of "tree reduction" shown in FIGURES 1.42, 1.43, and 1.44. What is the analogous process for the tree of FIGURE 1.52?

A basic tool in studying ordered sets is the notion of an order isomorphism.

1.54 DEFINITION.

Let $(P, \leq)$ and $(Q, \propto)$ be posets and let $\beta:P \to Q$ be a bijection. If $x \leq y$ in P if and only if $\beta(x) \propto \beta(y)$ in Q then P and Q are *order isomorphic* with order isomorphism β.

If the posets P and Q are both linearly ordered sets, then there is only one such order isomorphism (obtained by listing P and Q side by side and pairing off corresponding elements). For large sets this "listing and comparing" procedure is useless from both a computational as well as a conceptual point of view. Thus for important classes of sets that arise in combinatorics it is important to describe the order isomorphism β in a simple and natural way in terms of the structure of the sets involved. This problem is the subject of CHAPTER 3. We shall give a brief description of some of the basic ideas here.

1.55 DEFINITION.

Let S be a linearly ordered set with $|S| = s$. Let $\underline{s}$ denote the linearly ordered set $0, 1, \ldots, s-1$. We denote the order isomorphism between S and $\underline{s}$ by RANK and its inverse by UNRANK.

Thus, if $x \in S$ then RANK(x) is the number of elements of S that occur *before* x in the linear order on S. If $k \in \underline{s}$ then UNRANK(k) is the element of S that has exactly k elements preceding it. FIGURE 1.56 shows an edge $e = (a,b)$ in a tree structure such as those we have previously examined. The upper vertex of the edge is a and the lower vertex is b. Let $R(e) = R(a,b)$ denote the tree structure that consists of all paths of the form $a, x, \ldots$ and ending at a leaf of the tree where x is to the left of b. This tree structure is shown in FIGURE 1.56. We call $R(e)$ the *residual tree* of the edge e.

1.56 RESIDUAL TREE OF AN EDGE.

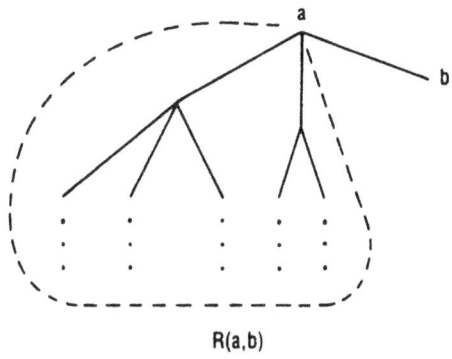

R(a,b)

Figure 1.56

In FIGURE 1.57, a tree structure is shown together with a path a_1, a_2, a_3, a_4 from the root to a leaf. The residual tree of each edge in the path is shown. If we let S denote the set of all leaves of the tree (equivalently, the set of all paths from the root to a leaf) linearly ordered from left to right, then the RANK(a_4) (equivalently, RANK(path a_1, a_2, a_3, a_4)) is just the sum of the number of leaves (or paths) in the residual tree of each edge of the path. If we let $\Delta(a_i,a_j)$ denote the number of leaves in $R(a_i,a_j)$, then RANK(a_4) = $\Delta(a_1,a_2)$ + $\Delta(a_2,a_3)$ + $\Delta(a_3,a_4)$ = 5 + 3 + 2 = 10. There are 10 leaves or terminal vertices before a_4 = k, namely, a,b,c,d,e,f,g,h,i, and j.

1.57 THE RANK OF A LEAF IN TERMS OF RESIDUAL TREES.

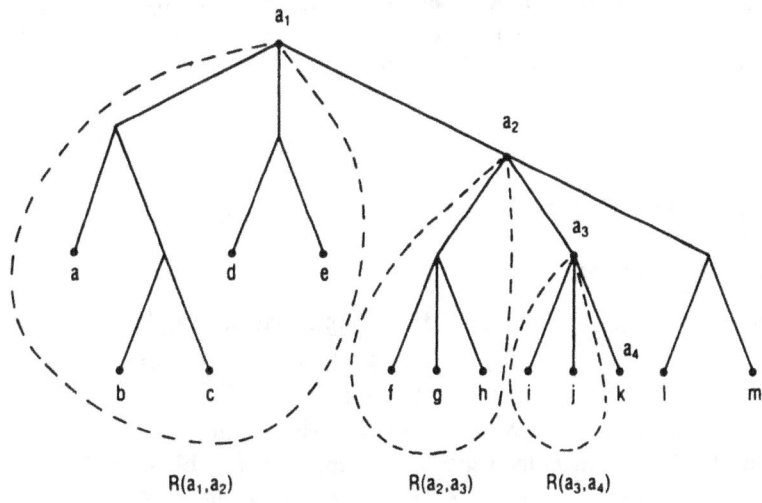

R(a_1,a_2)   R(a_2,a_3)   R(a_3,a_4)

Figure 1.57

33

1.58   RANK OF A LEAF OF D($\underline{6}^4$).

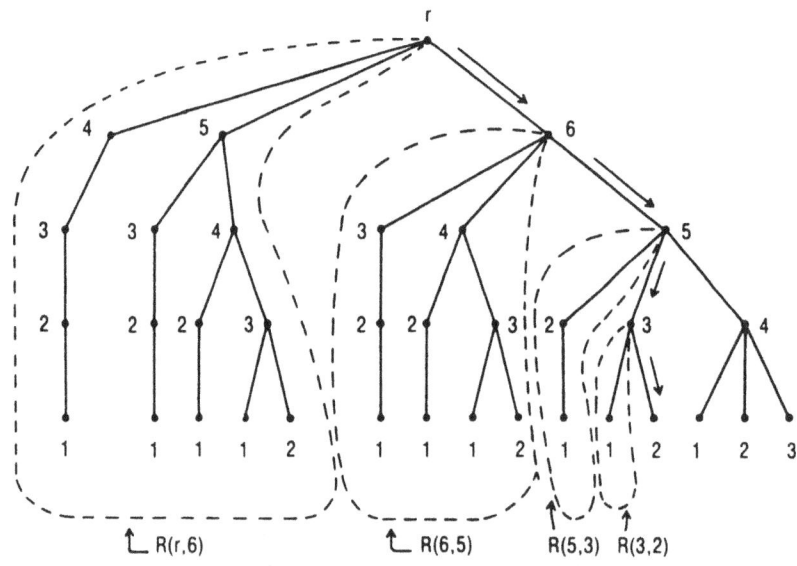

Figure 1.58

The idea illustrated in FIGURE 1.57 for computing the RANK of a leaf in a tree may be used to compute the RANK function of many important combinatorial lists. FIGURES 1.42, 1.45, 1.47, 1.48, and 1.52 are examples of basic lists that have been represented by tree structures. As an example, we show how to compute RANK and UNRANK of D($\underline{n}^k$) in lex order. FIGURE 1.58 shows the tree diagram of FIGURE 1.52 with path r, 6, 5, 3, 2 to a leaf. The residual trees of each edge in the path are shown by dotted lines. The RANK of the indicated leaf is thus Δ(r,6) + Δ(6,5) + Δ(5,3) + Δ(3,2). This sum is just

$$\binom{5}{4} + \binom{4}{3} + \binom{2}{2} + \binom{1}{1} = 11.$$ A moment's reflection on the general case

yields the following result.

1.59   THEOREM.

Let f be a function in D($\underline{n}^k$), the decreasing functions from $\underline{k}$ to $\underline{n}$, in lex order.

Then RANK(f) = $\binom{f(1)-1}{k} + \binom{f(2)-1}{k-1} + \ldots + \binom{f(k)-1}{1}$.

The computation of UNRANK is also based on the idea of a residual tree. Consider for a moment the general case represented by FIGURE 1.57. Suppose we wish to locate the leaf of RANK m. Let r denote the root and let $s_1, s_2, \ldots,$

34

s_p be the sons of the root in order from left to right. Which edge (r,s_i) must we take to eventually arrive at the leaf x with RANK(x) = m? Clearly, if $\Delta(r,s_i) > m$ then x lies in the residual tree $R(r,s_i)$, so we had better not descend along the edge (r,s_i). Also, if $\Delta(r,s_i) < m$ *and* $\Delta(r,s_{i+1}) \leq m$ then the largest (i.e., right-most) leaf in $R(r,s_{i+1})$ has RANK strictly less than m, so in this case also we would not want to descend along (r,s_i). Thus we must choose the index i such that $i = \max \{t: \Delta(r,s_t) \leq m\}$. We then let $m' = m - \Delta(r,s_i)$ and repeat the same procedure on the subtree rooted at s_i. This is the basis for the following result.

1.60 THEOREM.

Let UNRANK be the inverse of the function RANK of THEOREM 1.59. The following algorithm computes UNRANK:

procedure UNRANK(m) (Computes f = (f(1), . . . , f(k)) in $D(\underline{n}^k)$, RANK(f) = m)

    initialize $m' := m$, $t := 1$, $s := k$; $(1 \leq k \leq n, 0 \leq m \leq \binom{n}{k} - 1)$

    while $t \leq k$ *do*
        begin

$$f(t) - 1 := \max \left\{ y : \binom{y}{s} \leq m' \right\};$$

$$m' := m' - \binom{f(t) - 1}{s};$$

$$t := t + 1;$$

$$s := s - 1;$$

    end.

1.61 EXERCISE.

(1) Prove by induction that the procedure of THEOREM 1.60 is correct.
(2) Find the element UNRANK (99) of $D(\underline{10}^5)$ in lex order. Find RANK (10,8,6,5,2) in this list.
(3) State and prove the analogous results to THEOREM 1.59 and 1.60 for the list of permutations S_n in lex order and direct insertion order.
(4) Let S be a set (finite as usual). To "select an element at random from S" is to define a procedure that selects an element from S in such a way that any element of S is equally likely to be selected. In probability terms, the procedure selects an element from S according to the *uniform distribution* on S. Write procedures that select a permutation from S_n at random and a subset of size k from $\mathcal{P}_k(n)$ at random.

(5) How would you linearly order, rank, and unrank all ways of placing k indistinguishable balls into n bins? Try some examples first.

We put aside for now the computation of basic order isomorphisms for combinatorial lists. In CHAPTER 3 a detailed discussion of this subject will be given. The reader should note that if we have a nice algorithm for computing the order isomorphism RANK for a linearly ordered set, then, by computing RANK(x) where x is the last element of S, we also know the number of elements of S, $|S| = \text{RANK}(x) + 1$. The problem of counting the number of elements of a set S is a central problem of classical combinatorial theory (enumerative combinatorics). Thus the problem of computing RANK and UNRANK as posed above is in general more difficult than simply counting the total number of elements in the corresponding sets. In fact, many of the classical techniques for counting the number of elements of a set can be modified to produce methods for linearly ordering the elements of the set and computing RANK and UNRANK. This is particularly so for counting methods based on recursions as one can then use the partition tree concepts as previously discussed. Other methods with which the reader may be familiar (such as the principle of inclusion-exclusion) may not be readily modified to compute RANK and UNRANK. In some problems of listing combinatorial objects, the partition tree provides a useful framework but the trees do not have enough regularity to allow for a simple characterization of the residual trees. This situation is common in so-called backtracking problems ("forward leaping" problems, as noted previously). Problems involving both backtracking and isomorph rejections are especially complex and interesting in this regard.

We conclude our discussion with a famous example of this type of problem, the "8-queens problem." The great 19th century mathematician, Carl Friedrich Gauss, got the wrong answer to this problem! The classical "8-queens" problem is to discover all ways of placing 8 nonattacking queens on an 8×8 chessboard. A more general problem is to place n _or fewer_ nonattacking queens on an $n \times n$ chessboard. An $n \times n$ board has eight basic symmetries—the identity, rotation by 90, 180, and 270 degrees and reflections about the four axes ρ_1, ρ_2, ρ_3, ρ_4, as shown in FIGURE 1.62(a) for the case $n = 4$.

1.62   THE SYMMETRIES OF A CHESSBOARD.

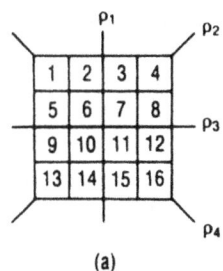

(a)

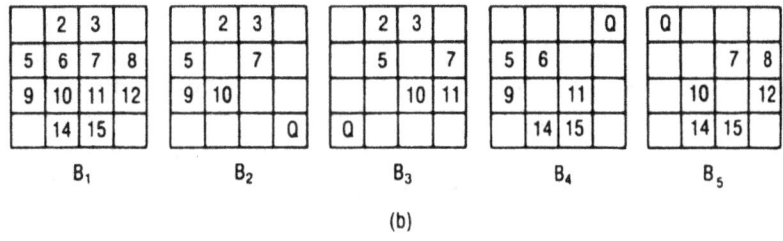

(b)

Figure 1.62

We number the squares of the board as one would read a book, from left to right and from top to bottom. Consider square 1 of FIGURE 1.62. By applying the group of symmetries G to the square, square 1 can be moved to the position occupied by square 4, 13, or 16, but to no other squares. The set $\{1, 4, 13, 16\}$ is called an *orbit* of G on the 4×4 board. This orbit could have been generated by rotations alone. An orbit that requires reflections as well as rotations is the set $\{2, 3, 5, 8, 9, 12, 14, 15\}$. There is one other orbit on the 4×4 board, $\{6, 7, 10, 11\}$. The orbits of a group of symmetries are always disjoint (they form a partition of the set of squares of the board). We linearly order the orbits according to the label of the *smallest* integer in the orbit. In FIGURE 1.62, B_1 shows the 4×4 board with the squares of the first orbit blanked out. In B_2 a queen has been placed on the S.E. (southeast) corner square. All squares attacked by this queen are also blanked out. Let Q_n denote the set of all solutions to the n or fewer queens problem and let $Q_n^=$ denote the set of all solutions to the exactly n queens problem. We regard the picture B_1 of FIGURE 1.62 as a symbolic representation of the subset of Q_4 consisting of all solutions that, when restricted to the first orbit (the corner squares), looks exactly like B_1 (i.e., the corner squares are empty). Likewise, B_2 stands for the subset of Q_4 consisting of all solutions with a queen in the S.E. corner and all other corners empty, etc.

37

We call the B_i "keys" or symbolic representations of certain subsets of solutions. We could equally well regard the B_i as specifying subsets of $Q\overline{\overline{4}}$. In this case, the set specified by B_1 has exactly two elements and the other B_i specify the empty set. We shall use keys such as the B_i to construct partition trees for Q_n or $Q\overline{\overline{n}}$.

1.63 EXERCISE.

Prove that there are only two solutions to the exactly 4-queens problem (they are isomorphic, as one can be obtained from the other by reflection ρ_1 of FIGURE 1.62).

The keys of FIGURE 1.62 are specified by giving an ordered subset S of the board ($S = \{1, 4, 13, 16\}$) and then a function from S to the ordered set $\square$, $\boxed{Q}$ (($\overset{1}{\square} \overset{4}{\boxed{Q}} \overset{13}{\square} \overset{16}{\square}$) for example). Assume for the moment that the functions on S are ordered lexicographically. We only use functions that represent nonattacking configurations of queens on the board. We now discuss a method for constructing keys and refining partitions for the queens problem. The interesting, and somewhat unusual, feature of the resulting partition tree is the way the orbits of the various symmetry groups are built into the tree structure.

1.64 EXAMPLE.

Consider the 4×4 board. The first (in lexicographic order) *orbit* of the full group of eight symmetries of the board is the ordered set $\{1, 4, 13, 16\}$ (remember, we order the collection of orbits *by their smallest element*). The four keys for this orbit are shown in FIGURE 1.62. Call the associated sets of solutions B_1, ..., B_5. For this example assume "solution" means a configuration of four or fewer queens on the board (i.e., Q_4). Suppose we select B_2. How do we refine B_2 into an ordered partition? A natural way that is recursive with respect to symmetry is to construct the next collection of keys on the first orbit with respect to the group of symmetries that leave the key B_2 unchanged. This group consists of the identity and a reflection about the line through the N.W. and S.E. corner of the board. The first orbit (with respect to this reflection group) in the region of the board that is not attacked by the queen is $\{2, 5\}$. The three keys for this orbit, B_{21}, B_{22}, B_{23}, are shown in FIGURE 1.65.

1.65   KEYS FOR REFINING ORBIT.

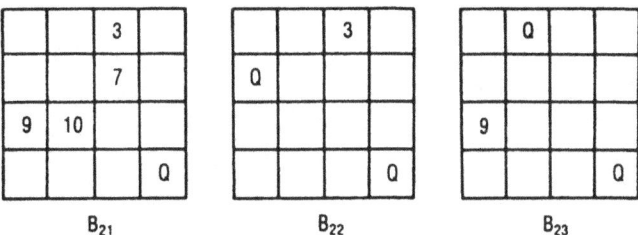

Figure 1.65

The set B_{21} again is fixed by the same group of order 2. The first orbit is $\{3, 9\}$, which gives rise to B_{211}, B_{212}, and B_{213}, shown in FIGURE 1.66.

1.66   THIRD ORDER REFINEMENT OF ORBIT.

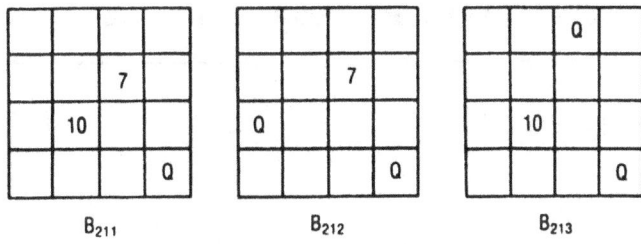

Figure 1.66

The set B_{211} again is fixed by the same group. The first (and last!) orbit is $\{7, 10\}$. This gives rise to B_{2111}, B_{2112}, B_{2113} (each consisting of one solution), as shown in FIGURE 1.67.

1.67   FOURTH ORDER REFINEMENT OF ORBIT.

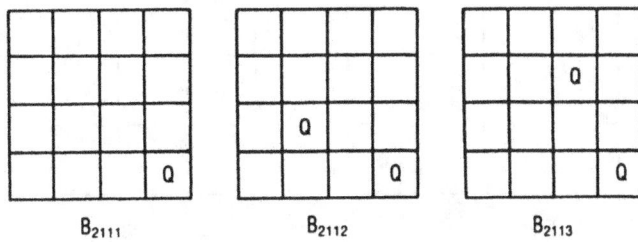

Figure 1.67

Discarding the B's and keeping their subscripting strings, the partition tree rooted at B_2 is shown by the tree diagram in FIGURE 1.68.

1.68   ORBIT PARTITION TREE.

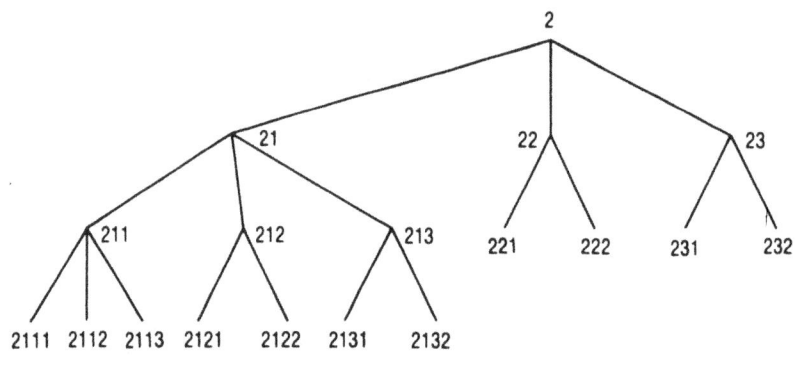

Figure 1.68

Some sample solutions are shown in FIGURE 1.69.

1.69   SOME SOLUTIONS TO 4- OR FEWER QUEENS PROBLEM.

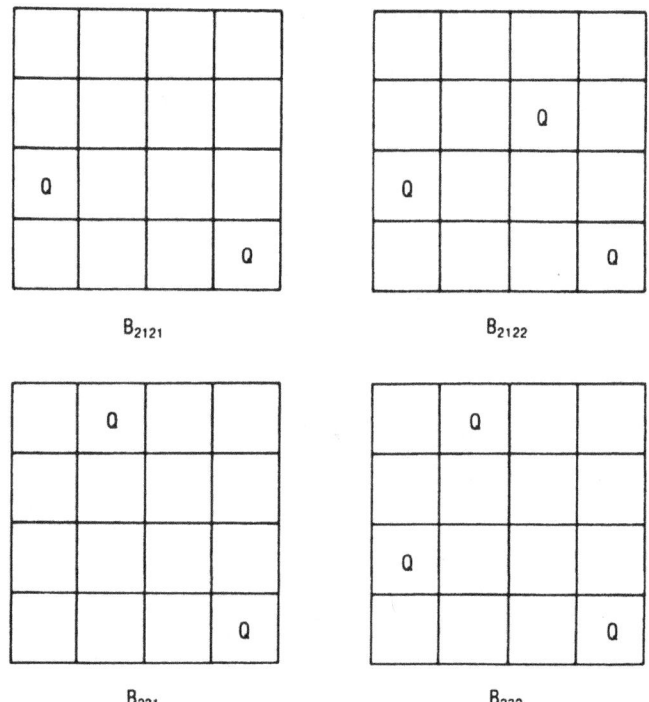

Figure 1.69

1.70 EXERCISE.

As in FIGURE 1.69, draw the solutions corresponding to the rest of the terminal nodes in FIGURE 1.68.

1.71 EXERCISE.

A direct way (not regarding symmetry recursively) to organize the n or fewer queens problem is column by column. A configuration of queens can be regarded as functions in $(\underline{n} \cup \{*\})^{\underline{n}}$. For n = 4 the configurations of FIGURE 1.69 would be specified by $2 * * 1$, $2 * 3\ 1$, $* 4 * 1$, and $2\ 4 * 1$. The numbers refer to height up the corresponding column. A $*$ means empty column. Carry out enough of this method of generating solutions to the 4- or fewer queens problem to get a feel for what is happening. What shortcuts can you make if you use this method on the "exactly n queens" problem?

1.72 EXERCISE.

Write a computer program to find all (there are 92) solutions to the 8- queens problem, $Q_8^=$. Can you also reject isomorphs (there are 12 solutions up to symmetries)?

Two solutions to the 8-queens problem are shown in FIGURE 1.73.

1.73 TWO SOLUTIONS TO 8-QUEENS PROBLEM.

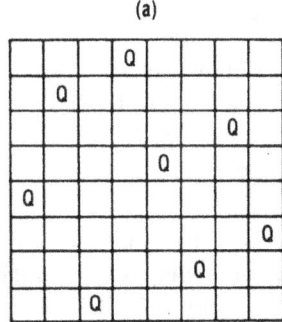 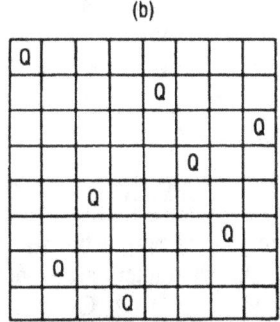

Figure 1.73

We use the symmetry-recursive ordering as in EXAMPLE 1.64. Consider 1.73(a). The first orbit under the symmetry group of order 8 of the board is $\{1, 8, 57, 64\}$. The first function ☐, ☐ ☐, ☐ is on this orbit, so write down B_1. The whole group fixes this function (i.e., leaves the key B_1 unchanged). The first orbit in the complement of the orbit and the region attacked by the queen configuration on the orbit (the latter, empty in this case) is $\{2, 7, 9, 16,$

41

49, 56, 58, 63}. The configuration on this orbit is again the first function, so write down B_{11}. The group fixing B_{11} is still the full group G. The first orbit in what remains is {3, 6, 17, 24, 41, 48, 59, 62}. The configuration □ □ □ □ □ $\boxed{Q}$ $\boxed{Q}$ □ is the 5th (nonattacking) configuration in lex order. Write down B_{115}. The group that fixes B_{115} is the identity. From now on orbits consist of single squares on the board. The first orbit in what remains is {4}. The configuration $\boxed{Q}$ is the 2nd (of two possible). Write down B_{1152}. The solution of 1.73(a) is thus contained in the set of solutions B_{1152}. Continuing this process we find that 1.73(a) is the one element in the set $B_{115222222}$.

1.74 EXERCISE.

Duplicate on 1.73(b) the procedure just employed on 1.73(a) to find the key (corresponding to 1 1 5 2 2 2 2 2 2 for 1.73(a)) that identifies 1.73(b). Do you see any pattern emerging on the keys that might help in finding solutions?

In Chapter 4 we shall study a class of algorithms called *orderly algorithms*, which may be used to deal with certain classes of *isomorph rejection* problems. The isomorph rejection problem for Q_n would be to obtain a subset Δ_n of Q_n such that no two elements of Δ_n can be transformed from one to the other by rotations and reflections of the board (they are inequivalent with respect to the group of symmetries of the board) but every element of Q_n can be obtained from some element of Δ_n by a rotation or reflection of the board (Δ_n is complete relative to Q_n). The set Δ_n is called a *system of representatives* of Q_n with respect to the group of symmetries of Q_n. One approach to this problem is to first generate all of Q_n and then select Δ_n. This approach was taken in connection with the domino covering problem of FIGURE 1.37. The *"symmetry recursive"* partition tree that we have been discussing above allows for a recursive approach to the isomorph rejection problem. We illustrate the basic ideas with the problem of generating a representative system Δ_5 for Q_5.

1.75 GENERATING Δ_5 for Q_5.

We begin by numbering the squares of the 5 × 5 chessboard from 1 to 25 in the order that characters are printed on a page, just as in the case of the 4 × 4 board of FIGURE 1.62. Orbits are numbered as in FIGURE 1.76.

Take the empty square bearing the smallest number, in this case 1, and compute its orbit under the full symmetry group of the square—in this case, squares 1, 5, 21, and 25. The largest number of queens that may be placed in these four squares is one; the other possibility is zero. We begin by placing one queen on square 1. By the use of the symmetry group, we need not consider placing one queen on square 5, square 21, or square 25; this amounts to picking a representative of an orbit for the symmetry group acting on the configuration of queens on these four squares. Later we will pick another representative of an orbit under this action. We will keep track of these choices by always placing the "largest" first. That is, if the presence or absence of a queen on a square is interpreted as a binary digit, square 1 is the most significant bit, and square 25 is the least significant bit. A record must be kept of the orbits as they are used. *This represents a variation on the linear order that results on Q_5 as compared with the order on Q_4 discussed previously.*

From the recursive viewpoint, the problem is now the following: place four or fewer nonattacking queens on the 21 squares that are exclusive of squares 1, 5, 21, and 25 so that they are not attacked by the queen in square 1. Note there is to be a queen in square 1, and no queen in squares 5, 21, and 25. The symmetry group of this problem is the group of order two consisting of the identity and reflection in the diagonal through squares 1 and 25. The orbit in the chessboard containing the least numbered unspecified square consists of squares 2 and 6. The "largest" placement of queens on this orbit is to place no queen on either square 2 or square 6.

The algorithm now continues in this recursive fashion. As soon as the identity group is reached, isomorphs are rejected, and the search may proceed in any manner. When orbits of various groups acting on the chessboard fill up all 25 squares, a solution has been reached. One then finds the largest orbit (in the sequence of orbits of the various symmetry groups) for which the configuration of queens on that orbit can be advanced. "Advanced" means changed to the next orbit representative for the symmetry group of that orbit acting on the configuration of queens in that orbit.

An initial and a terminal segment of this algorithm is executed step by step in FIGURES 1.76 and 1.77. Orbits are numbered as they are set down; queens are denoted by circles. For the sake of brevity, only those steps in the execution of the algorithm that change the configuration of queens, are a backtrack, or are terminal are recorded. Terminal steps (when the whole board is filled by labeled orbits) are solutions.

1.76   INITIAL SEGMENT IN THE GENERATION of Δ₅.

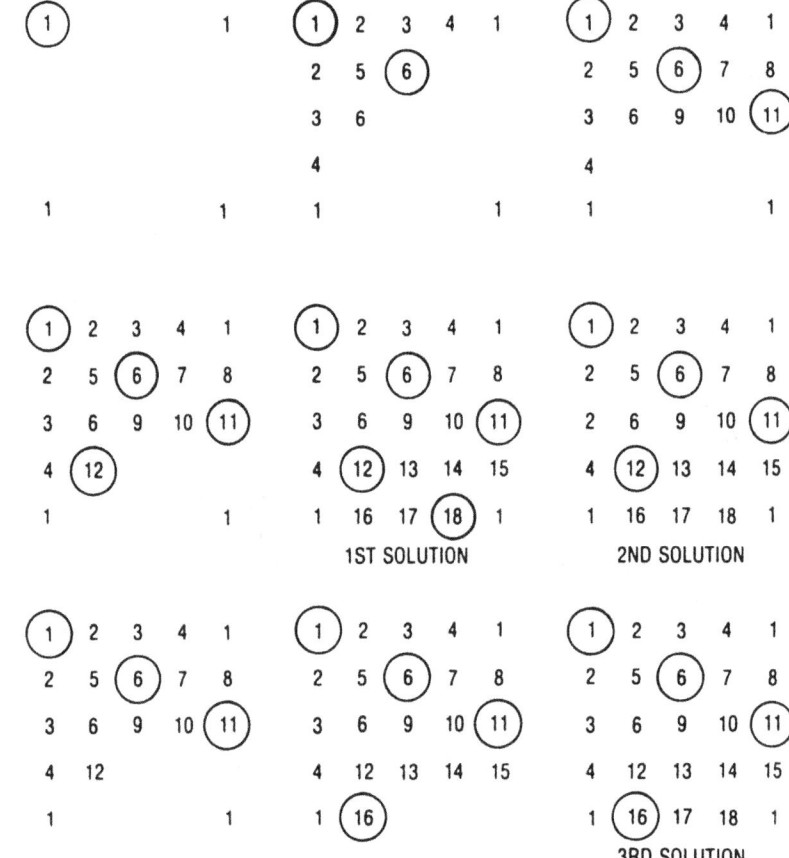

Figure 1.76

44

1.77 TERMINAL SEGMENT IN THE GENERATION OF Δ_5.

```
1  (2)  3   2   1        1   2  (3)  2   1        1   2  (3)  2   1
2   4   5   6   2        2               2        2   4   5   4   2
7   8   9  10  11        3               3        3  (6)      6   3
2  12  13  14   2        2               2        2               2
1   2  15   2   1        1   2   3   2   1        1   2   3   2   1
       55TH SOLUTION


1   2  (3)  2   1        1   2  (3)  2   1        1   3  (3)  2   1
2   4   5   4   2        2   4   5   4   2        2   4   5   4   2
3  (6)  7   6   3        3  (6)  7   6   3        3   6       6   3
2   8   9  (10)  2       2   8   9  10   2        2               2
1   2   3   2   1        1   2   3   2   1        1   2   3   2   1
       56TH SOLUTION            57TH SOLUTION


1   2  (3)  2   1        1   2  (3)  2   1        1   2  (3)  2   1
2   4   5   4   2        2   4   5   4   2        2   4   5   4   2
2   6   7   6   3        3   6   7   6   3        3   6   7   6   3
2  (8)      8   2        2  (8)  9   8   2        2   8       8   2
1   2   3   2   1        1   2   3   2   1        1   2   3   2   1
                               58TH SOLUTION


1   2  (3)  2   1        1   2   3   2   1        1   2   3   2   1
2   4   5   4   2        2               2        2  (4)      4   2
2   6   7   6   3        3               3        3   3       3
2   8   9   8   2        2               2        2   4       4   2
1   2   3   2   1        1   2   3   2   1        1   2   3   2   1
       59TH SOLUTION
```

Figure 1.77

(cont.)

45

```
1   2   3   2   1        1   2   3   2   1        1   2   3   2   1
2  (4)  5   4   2        2  (4)  5   4   2        2   4       4   2
3   5   6  (7)  3        3   5   6   7   3        3               3
2   4   7   4   2        2   4   7   4   2        2   4       4   2
1   2   3   2   1        1   2   3   2   1        1   2   3   2   1

    60TH SOLUTION            61ST SOLUTION


1   2   3   2   1        1   2   3   2   1        1   2   3   2   1
2   4  (5)  4   2        2   4  (5)  4   2        2   4   5   4   2
3   5       5   3        3   5   6   5   3        3   5       5   3
2   4   5   4   2        2   4   5   4   2        2   4   5   4   2
1   2   3   2   1        1   2   3   2   1        1   2   3   2   1

                            62ND SOLUTION


1   2   3   2   1        1   2   3   2   1
2   4   5   4   2        2   4   5   4   2
3   5  (6)  5   3        3   5   6   5   3
2   4   5   4   2        2   4   5   4   2
1   2   3   2   1        1   2   3   2   1

    63RD SOLUTION            64TH SOLUTION
```

Figure 1.77 (continued)

46

Index

48

NOTES

NOTES

NOTES

NOTES

www.ingramcontent.com/pod-product-compliance
Lightning Source LLC
Chambersburg PA
CBHW081243180526
45171CB00005B/523